Online Searching Technique and Management

Edited by James J. Maloney

American Library Association · *Chicago 1983*

Cover design by Ray Machura

Text composed by FM Typesetting Company in Linotype Caledonia. Display type, Caslon, composed by Pearson Typographers.

Printed on 50-pound Glatfelter, a pH-neutral stock, and bound in 10-point Carolina cover stock by Thomson-Shore, Inc.

Library of Congress Cataloging in Publication Data
Main entry under title:

Online searching technique and management.

Includes bibliographies.
1. On-line bibliographic searching—Congresses.
2. Machine-readable bibliographic data—Congresses.
3. Bibliographical services—Congresses. 4. Reference services (Libraries)—Congresses. I. Maloney, James J.
Z699.A1055 1983   025.3′028′54   83-11954
ISBN 0-8389-3285-1

Copyright © 1983 by the American Library Association. All rights reserved except those which may be granted by Sections 107 and 108 of the Copyright Revision Act of 1976. Printed in the United States of America.

Contents

Preface	v
PART 1. TECHNIQUE	
Online Searching: Past, Present, and Future SARA D. KNAPP	3
Online Searching in Perspective: Advantages and Limitations JANE I. THESING	16
Initial Considerations for an Online Search CAROL M. TOBIN	22
Databases and Database Producers and Vendors CAROL HANSEN FENICHEL	26
Physical Requirements: Terminals, Printers, and Furniture JANET BRUMAN	35
The Interview Process in Online Searching LAWRENCE R. MAXTED	50
The Mechanics of Online Searching KRISTINE SALOMON	57
PART 2. MANAGEMENT	
Library Organizational Patterns in Online Retrieval Services PETER G. WATSON	75

Contents

Who Should Search?: The Attributes of a Good Searcher 83
RANDOLPH E. HOCK

Library Applications of Database Searching 89
M. R. DUSTIN

The Policy Manual for Online Reference Services 103
GERTRUDE FOREMAN

Forms and Record Keeping for Online Searching 107
GAYLE MCKINNEY

Costs, Budgets, and Financial Management 123
NANCY E. GRIMES

Methods of Funding 135
JOHN EDWARD EVANS

Options in Training and Continuing Education 149
JAMES RETTIG

Publicizing an Online Search Service 160
MARYJANE S. COCHRANE

The Impact of Online Search Services on Libraries 166
REBECCA WHITAKER

APPENDIX

An Introduction to Online Searching: A Suggested Outline 173
MARS EDUCATION AND TRAINING OF SEARCH ANALYSTS COMMITTEE

BIBLIOGRAPHIES

Online Reference Service: How to Begin 181
EDITED BY EMELIE J. SHRODER

The Evaluation of Information Retrieval Services: A Selected Bibliography 189
COMPILED BY RASD-MARS COMMITTEE ON MEASUREMENT AND EVALUATION OF SERVICE

CONTRIBUTORS 193

Preface

These papers are based on the proceedings of a day-long program of the Machine-Assisted Reference Section (MARS) of RASD, held on July 11, 1982, at the annual conference of the American Library Association in Philadelphia. In two separate sessions the program addressed the needs of the novice searcher and of the manager of an online search service. The morning portion of the program was devoted to an introduction to online searching for the novice searcher. Papers discussing the management of an online search service were delivered in the afternoon.

The program resulted from a decision of the MARS Executive Committee to respond to librarians who attended MARS committee meetings seeking advice or guidance on a variety of recurring questions, such as:

> What is the relation between database searching and the more traditional library services?
> When does one decide to conduct an online search?
> How does one choose a database vendor?
> What is database searching used for?
> How does one resolve the unique financial considerations prompted by database searching?
> How can an online search service be funded?
> What are the characteristics of a good search analyst?
> What preparation and training are necessary for the professional librarian to become a search analyst?

The question What is database searching used for? is asked not only by novice searchers and librarians initiating online database access, but also by librarians seeking innovative users for an established online search service.

Questions of this nature are the focus of considerable attention and effort on the part of committees within MARS. These committees regularly gather information in order to study issues or problems as they affect database searching, to make recommendations or establish guidelines, or to serve as advocates on behalf of the community of database searchers. The framework for the 1982 MARS program was provided by one of those committees, the Education and Training of Search Analysts (or MARS/ETSA) Committee. Under chairs Greg Byerly (1981) and Rebecca Whitaker (1982), the MARS/ETSA Committee developed "An Introduction to Online Searching: A Suggested Outline" (included here as the appendix). This outline served as a checklist for librarians by profiling the basic concepts of database searching and elaborating the applications of this service to library work. The work of the MARS/ETSA Committee served as an ideal starting point for the members of the 1982 MARS Program Committee to select an outline of topics for the program, since the program would address the questions that were most frequently asked regarding the basic concepts and applications of database searching.

With three exceptions, all papers for the program were chosen by the Program Committee during the 1982 ALA Midwinter Meeting from nearly one hundred abstracts that had been submitted in response to a general "call for papers" in the fall of 1981. Because of the wealth of their experience, wit, and insight into the field of database searching, Sara Knapp and Peter Watson were approached by the committee to deliver keynote addresses for the morning and afternoon sessions, respectively. Carol Fenichel volunteered on the basis of her extensive teaching and writing experience in the field of database searching to fill any gap that the program committee might have. The committee responded by having her address the difficult topics of databases and database producers and of database vendors and search services.

This work addresses the basic concepts and applications of database searching, the various institutional considerations such as policies, procedures, record keeping, and promotion as they relate to a search service, financing, and the training of database searchers.

As a text on database searching, the work is a unique publication insofar as it is drawn from the experience, background, and perspective of a variety of people. Sally Knapp's view on how the public and private sectors should determine future development of online searching reflects the convictions and principles of a librarian who has actively engaged and shaped this issue. Randolph Hock's chapter, "Who Should Search?," is a thorough and emphatic analysis of the attributes of a good searcher, from someone who has personally trained over 2000 searchers. To quote Hock, "I feel strongly that the future of reference librarianship is going to be determined to a large degree by the use of online systems, that the quality of reference service will largely depend upon the quality of searching, and that the quality of searching will be dependent upon who is searching." Carol Fenichel's paper exudes with confidence that "soon it will be feasible to do all reference work online." The selection of equipment for online searching is expressed by Janet Bruman with the clarity and detail of someone with considerable professional experience.

This work also represents the contributions of many librarians who have actively participated in the work of MARS committees. In addition to the contribution of the MARS Education and Training of Search Analysts Committee previously referred to, much of the work of MARS Cost and Financing Committee is embodied in the paper presented by Nancy Grimes, the chair of that committee, entitled "Online Reference Services: Costs, Budgets, and Financial Management." The essence of a work based on the 1982 MARS program lent itself well to the inclusion of the two separate, selected bibliographies of materials recently compiled by the MARS Committee on Measurement and Evaluation of Service and the MARS Use of Machine-Assisted Reference Services in Public Libraries Committee.

Credit for the program is given to the members of the 1982 MARS Program Committee whose energy and enthusiasm for the project made the program possible. The committee members included:

Nathan Baum, Reference Department, Melville Library, State University of New York Stony Brook, New York

Linda Friend, Patee Library, Pennsylvania State University, University Park, Pennsylvania

Rachel Gade, Ramsey County Public Library, Roseville, Minnesota

Bernard F. Pasqualini, Free Library of Philadelphia, Philadelphia, Pennsylvania

Maria J. Soule, Learning Resource Center, Florida Keys Community College, Key West, Florida

Credit is also given to the MARS Executive Board and, in particular, to Guy T. Westmoreland and to Pamela C. Sieving, for the inspiration which they provided for the program. Special thanks is also extended to Andrew Hansen, RASD Executive Director, for the patience, practical advice, and effort which he extended on the committee's behalf.

JAMES J. MALONEY

Part 1
TECHNIQUE

SARA D. KNAPP

Online Searching: Past, Present, and Future

Definitions

Online searching has been a subject of interest to librarians for over a decade. In 1982, literally millions of searches were conducted and the ranks of online searchers continue to grow. What is online searching?

Online searching has been defined as a means of finding desired information, usually bibliographic references, by using a computer in an interactive mode.

A *database* has been defined as a collection of information called machine-readable records. In our work, this often corresponds, in content, to a printed index.

Every day hundreds of searchers independently but simultaneously search the vast array of databases made available to them online by producers and vendors in this country and abroad.

Growth of Searches and Databases

The growth of online searching has been remarkable and continues to climb. In 1974, Martha Williams reported that 700,000 online retrospective searches were conducted via online services in the United States and Canada. In 1976, 1,200,000 were reported and the figure had jumped to two million by 1977. That figure had doubled again for an estimated four million searches by 1979.[1] In 1982, Williams reported that the rate of growth of online use is declining slightly "but in absolute numbers the growth continues."[2]

Paralleling the growth in the number of searches performed was the increase in the number of available databases. Martha Williams's *Computer Readable Data Bases* listed 301 U.S. databases in 1975, 528 by 1979, and the 1982 edition lists over 750 available for public access, worldwide.[3]

3

History

The origins of the computer go back two centuries to Leibnitz's ideas for a calculating machine and, more recently, to Babbage's arithmetical precursors, Hollerith cards, otherwise known as punch cards, first used in 1890 for the U.S. Census. During World War II, computers were used secretly for cryptography.

As with many other aspects of our present life, the potential of computers began to be realized in that great surge of development following World War II. The application of computers to handling large amounts of diverse information in novel ways struck the imagination of many, and some accounts foreseeing direct use of large banks of data by untrained people made their way into popular and professional literature.[4]

One of the applications of computers during the early 1960s, computer-assisted typesetting, led to the development of the online databases of today. Publishers of large information services, such as those which abstracted journals, discovered that computer-assisted typesetting had a useful by-product. The information contained in the publication was in a form which a computer could read and manipulate. Publishers of these information services had indirectly created a database through this method of typesetting at virtually no additional expense or effort over that involved in normal production of a printed index.[5]

These databases were made available to the public during the early 1960s through the resources, generally, of university computer or information centers. These centers conducted searches of all or specific portions of databases that were loaded onto the center's own computer. The centers also conducted current-awareness searches, in which new material that was added to a database at periodic intervals was searched against the information profile of a client researcher. These current-awareness searches are commonly referred to as *selective dissemination of information* or *SDI* searches.

Under the present method of conducting a database search, the librarian interacts directly with the database through a computer terminal. The results of the search can be known immediately. In contrast, clients of computer or information centers of the early 1960s would often wait for days or perhaps weeks after submitting a request to see the results of the search, because the center batched, or combined, several requests in order to reduce the operating cost of each search.

Several early services were operational in batch mode. In 1964, the National Library of Medicine (NLM) began offering a batch service on requests that had been referred to it.[6] In batch-mode searching, an entire search request is submitted, neither too broad nor too narrow, and then the searcher waits two weeks for the search results. Batch searching is rather like playing the blindfolded game of pin-the-tail-on-the-donkey; you make what you hope is a stab in the right direction, but you lack the feedback to know how to adjust your strategy.

Actually, online searching was first investigated in the 1950s and was first publicly demonstrated by the Systems Development Corporation (SDC), in 1960, but that system lacked the ability to use Boolean operators or to back-reference previous search statements. SDC began development of the precursor of its retrieval language, ORBIT, in the mid-1960s. This also was later improved and used in NLM's Elhill system for MEDLINE which in the 1969-71 period was operated by SDC.[7]

Lockheed began in 1965 to develop its DIALOG system to provide an online system for NASD and began regular searches services by 1967.[8]

In 1968, the State University of New York (SUNY) Biomedical Communication Network became operational, offering online access to MEDLARS, NLM Current Catalog tapes and a database of medical library monograph holdings. The network used an early system, IBM Document Processing.[9]

Since I learned searching on that old, rather primitive system, I would like to share some recollections of what searching was like then. First of all, the terminals were hard-wired to the computer. No dial-ups, you just left your terminal on all day. (Every now and then it would type a message such as that the system would be down tomorrow or that the SUNY Central Staff wished you a Merry Christmas.)

The system allowed you to back-reference search statements but you had to put in your entire strategy before you got any results. To change it you had to re-enter the entire strategy. You couldn't print descriptions for citations so you couldn't always tell which part of your strategy was causing problems.

We had an old clunker of a terminal. It was noisy and printed only a 110 baud (10 characters per second), which we thought was just fine. Response time of a minute or two didn't seem so bad, either.

In reminiscing with other searchers I have wondered which of our expectations of that period had or had not been fulfilled and how the outcomes of those expectations that might help us look at the predictions of today.

I know some of us expected online searching to grow far faster than it did. It seemed so obviously wonderful to me that I was amazed at the lack of interest and even opposition to it among some contemporary librarians. I think one can say that change is more likely when those responsible for it perceive it as being in their own interests. Many librarians and administrators did not so perceive online searching.

Another librarian told me she had thought we'd be able to integrate searching into local operations, like being able to have call numbers on searches and run them against a local holdings database. Integrated searching operations, too, have waited on technology and incentive for development.

Others of us hoped for standardization among database elements. Little has been done in this area, probably because until now many producers have seen little value in it. Some vendors who recognize users' needs have compensated for the failure of producers to standardize.

But a number of changes were necessary for the evolution of our present systems. One of the most significant was the development of inexpensive dial-up communications linking remote terminals to the large computers of the online vendors.

Packet Switching and Telecommunications Networks

This development of packet switching and telecommunications networks now makes possible worldwide online access via networks such as Tymnet and Telenet. Being hard-wired to the old SUNY BCN (Biomedical Communication Network) computer meant that one line from the computer was dedicated to us and our terminal was dedicated to it. It limited the number of users who could be accommodated.

Tymnet, first offered by Tymshare, Inc., in 1969, was the first widely available data network which made possible the present flexible and more economical mode of communication. "Based on 'packet switching' these systems divide the input flow of information into small segments or packets of data which move through the network in a manner similar to the handling of mail but at immensely higher speeds."[10] By so doing, the same lines can be used for many communications from many widely dispersed users at the same time. The Tymnet network was based on minicomputers which linked various terminals to central computers. The network switches, which were linked by voice-grade telephone-type lines, could store and then forward the data from node to node, using the fastest route to a destination.[11] Today, the networks—Tymnet, Telenet, Uninet, and others, too—translate signals from all kinds of dissimilar terminals.[12]

Development of National Search Services

The emergence of packet switching and telecommunications networks made possible the further development and expansion of the national search services which we know today.

Before the advent of Tymnet in 1969, accessibility to online databases was limited. Lockheed had developed DIALOG, which made a handful of scientific and technical databases available to the National Aeronautics and Space Administration (NASA). A version of DIALOG had also been developed by Lockheed for the European Space Research Organization, which was later used by the European Information Retrieval System (IRS). SDC's ORBIT system was available to the U.S. Air Force over telephone lines. The SUNY BCN was providing medical libraries with access to databases created by the National Library of Medicine.

By 1971, NLM's MEDLINE had become operational as a commercially available database. In 1972, both DIALOG and SDC began to enlarge the small selection of their scientific and technical databases, by making their services available to a wider market of users. BCN, which was the predecessor to Bibliographic Retrieval Services (BRS), ceased in May 1977 when BRS became operational.

Of the national search services referred to, DIALOG, SDC, and BRS are commercial vendors of databases. Database vendors contract with the producers of databases, such as the National Library of Medicine or the Educational Resources Information Center (ERIC), to make the databases which they create available through their systems. A vendor system encompasses the computer equipment used to store multiple databases, the staff of technical and public services personnel employed to maintain that equipment and to facilitate its use for the public, and the software developed to enable users to search those databases.

Database producers are the federally supported institutions, agencies of the federal government, professional associations, and corporations that produce the machine-readable records that usually correspond to a printed index. Most database producers simply make their machine-readable product available to the public through one or more of the database vendors. Some database producers such as the National Library of Medicine, however, have made their databases available through their own vendor systems, as well as those of the commercial vendors. This marketing practice has worked well for a producer such as NLM which, in response to the information needs of the medical community, has developed several different databases sufficient to warrant maintaining a separate vendor system. There is a trend among database producers, which I will discuss later, to develop their own vendor systems which allow them to market their databases directly to the public.

The momentum of developments in database searching during the 1970s continued with the emergence of online users' groups in the mid-1970s. In 1976, requests to join RASD's Information Retrieval Committee were so numerous that the chair, Peter Watson, proposed forming a

8   Technique

Machine-Assisted Reference Services Discussion Group, and so MARS was born at the ALA Conference held in Chicago that year. By 1978, MARS had so many members with so many interests that it was able to become an RASD section with its own committees. Two new publications, *Online* and *Online Review*, emerged in 1977, followed by *Database* in 1978. In 1982, *RQ* began carrying reviews of databases.

Recently we have witnessed the widespread conversion to 1200 baud, a development that has reduced connect time for searching. Conversion to 1200 baud has led to changes in pricing by database producers. Many of them are adding "per hit" charges, which means that the searcher pays whenever he or she finds something in the database, as well as for the time connected.

The Future

At least one expert predicts that in the next ten to fifteen years major changes in the way we interact with computers will be possible. These include natural language capabilities, speech understanding, and speech generation. In addition, computing costs will be low enough to make these features feasible.[13]

Another expert predicts an end to offline printing, with new pricing schemes to accommodate that change and the documentation of search results by microcomputers for subsequent printing and manipulation.[14]

End Users and Information Professionals: Changing Roles

Will our patrons learn to search for themselves? What will this do to us? There is absolutely no doubt that direct searching by users is in the offing. Already, user-friendly interfaces have been developed. Easy systems using lists of choices and simple commands are available to home users, and some are being marketed to libraries as well. Admittedly these are not yet very sophisticated or powerful systems, but far greater possibilities are now being developed. Systems linked with full text databases, which will allow users to browse easily, can literally put a library of information at one's fingertips. Using the kind of associational trails dreamed of by Bush and others since the 1940s, users will be able to follow their own logic and their own associations to the exact information sought.

The users will be ready for them. Some of our generation may have faced terminals with trepidation, but that will certainly not be the case with today's youngsters who use microcomputers in schools and computer games at home and who find terminals and computers quite a normal part of everyday life. One college president is predicting that within a few years college freshmen will be required to have their own microcomputers.[15] Even today's adults find user-friendly systems comfortable and enjoyable to use. The recent report of the "paper chase" self-service system indicated a fantastic response by users of a medical library.[16]

What of the role of libraries and librarians? Where will we be? Experts seem to differ. One well-known writer points out that, although the do-it-yourself trend is widespread, we frequently pay others to mow lawns, paint houses, repair our cars, and even do our taxes. Presumably there will always be those who want to have someone else do their searching. Others point out that searching is very personal. They foresee such sophisticated and yet simple-to-use systems taking users beyond citations to actual data with so much assistance that neither searchers nor libraries will be needed.

The difference between these views may just be a question of how far in the future you care to look. When we think of doing searches in the near future, we think in terms of present systems. It seems incredible that very many end users would want to learn the myriad conventions, codes, access points, and controlled vocabularies we now use in searching. User-friendly terminals and systems will be quite different and it is likely that as they evolve, more and more of the searches we now do will be more easily done at the user's own terminal and convenience. Nor will users have to wait for documents referred to by databases, since they will also be online. Present systems are indeed cumbersome, unwieldy, and demanding of the user.

In the immediate future, however, there are at least five roles for information professionals:

1. To continue to provide searches and service to those who are disinclined or unable to do them for themselves[17]
2. To do the difficult, unusual, or offbeat searches not so readily handled by the first generation of user-friendly systems. An important part of this role may be in assisting users to analyze their search requests and in helping them to think about their questions[18]
3. To educate users about available information services and help them learn to use them[19]
4. To act as advocate for the public, especially our own user communities, in identifying and working to obtain the kinds of services needed that do not exist and in evaluating the effectiveness of those that are in existence[20]
5. Another role, suggested by Information Industry Association (IIA) President Paul Zurkowski, is to be a sort of coordinator of locally produced databases.[21]

Producers and vendors have a tremendous interest in promoting end-user searching. When everyone has their own terminals, the market will be almost unlimited. Intermediary searching can never promise the volume of business that end-user searching will provide. So there are tremendous economic incentives for bypassing intermediaries. And technologically, the combination of artificial intelligence and microcomputers as terminals promises to make searching very, very easy.

Private/Public Sector

The question of public or private funding of the production, distribution, and use of online information systems will surely have a profound effect on the future of online services. The economic argument over whether those who use a service should pay for it began with the issue of user fees in publicly supported libraries and now extends to the propriety of government as database vendor, for example in providing the MEDLINE network, or as the producer of such widely used databases as ERIC, AGRICOLA, MEDLARS, NTIS, and the National Institute of Mental Health.

Advocates range along a continuum from those who believe government has an obligation to provide certain kinds of information services to those who would merely refrain from restricting the government role to those who would carefully limit the role of government. There was disagreement on these fundamental issues between members of the National Commission on Libraries and Information Science Public Sector/Private Sector Task Force and I am sure there is disagreement among ALA members. The report of the NCLIS Task Force seems to have as one of its most important but controversial elements the encouragement of private enterprise in the development of information as a national resource. Reliance on the marketplace would make the criteria of value economic rather than political. Reducing the value of information solely to economic criteria is controversial: we might argue that unless everyone has the dollars to spend on information, economic factors aren't always a fair measure of the public's need.

But there is certainly much to be said for private enterprise as a means by which products are manufactured and distributed. It is private enterprise that put all sorts of appliances and products in our homes, and it is generally believed that profit encourages competition . . . which tends to lower prices and encourage the widest possible dissemination through marketing. Obviously this competitive element is going to apply to products and services for which there is demand by those who can afford to pay, in other words products and services which meet economic criteria.

It seems, however, to clash with the philosophy of the authors of the Library Bill of Rights, who wrote: "Libraries should provide books and other materials presenting all points of view concerning the problems and issues of our times. . . ." Could an unpopular view meet economic criteria? Commercial television seems to exemplify the lowest-common-denominator effect of applying only economic criteria to information dissemination. At any rate, the proposed use of the private sector for the public purposes of the creation and dissemination of information has led to the introduction of a bill, HR 4758, aimed at protecting the rights of the private sector against unfair competition by the government. With the backing of the Information Industry Association, this bill would prohibit federal agencies from providing data processing or telecommunications services to any user other than another federal agency unless it cannot be provided by private persons, in which case the bill would require full reimbursement for all costs including overhead.[22] IIA is proposing additional amendments requiring that "federal agencies charge for any service or product they are allowed to offer according to what it costs them to produce it." Proponents of the bill and its proposed amendments say that by not recovering costs, the government is in a position to drive out of business companies and nonprofit organizations and then to have an undue influence over research by becoming the sole supplier of information in a field.[23]

Estimates are that with full cost recovery a subscription to printed *Index Medicus* would go from $200 to $1000 and online access to MEDLARS from between $15 and $22 an hour to $73 an hour.[24] It is the Dutch Company, Elsevier, publishers of *Excerpta Medica*, who claim that NLM is not allowing them to make a substantial profit.[25] Should the right to make a profit take precedence over the right of citizens to require their own government to provide them with certain kinds of services? Are the facts being ignored that (1) any producer of a large database with an extensive back file has a virtual monopoly because of the incredible expense which would be required to duplicate that back file, and (2) many producers who hold such monopolistic positions did, in fact, develop their databases with federal grants? If this legislation passes and prices for what are now government-produced databases rise, I believe the tremendous growth in database searching will decline. I know that in our library government-produced bases accounted for 47 percent of the searches we processed last year. I'm sure our figures are not unrepresentative of other academic and medical libraries. I'm equally sure that, at twice the price, most of the searches would *not* have been performed and those information needs would have been unmet or poorly met.

If the volume of searching declines drastically, can private enterprise really be doing a better job than government in meeting the needs of the public?

Microcomputers

Microcomputers, which are becoming increasingly affordable, promise many improvements in online searching. Among these are the ability to store and edit search output; automatic searching; storage of profiles; storage of search aids; automatic dial-up and sign-on; simplified interfaces for the unskilled; translation from one system's commands to another; and simulation systems for training searchers cheaply.[26]

Publishers as Vendors

Another new trend is for producers of secondary databases to bring up their files on their own computers and to sell them directly, rather than through the major vendor services like DIALOG, BRS, and SDC. One producer wrote us a letter explaining its reasons and suggesting that we sign a contract. Since it was a database we used only a few times a year, we simply decided to forego using that file in the future. Libraries don't want to, and in fact can't, deal with a proliferation of systems and contracts. That is exactly why, for years, libraries have used jobbers rather than purchasing books and periodicals from individual publishers.

Admittedly, some files are so indispensable that we would have to have them wherever they might be found, but competitive files mounted on a big vendor's system would have a clear edge over independent ones. Files used occasionally by many libraries when offered by a vendor would simply not be searched if they were no longer on the vendor's system. The practice of a publisher selling direct seems a very misguided approach for abstract and index producers, considering that they themselves may be vulnerable to technological change. Electronic publishing of primary literature coupled with sophisticated searching systems including full-text searching may decrease the need for indexing and thus allow users to bypass abstracts and indexes.[27]

Nonbibliographic Databases

Nonbibliographic databases have seemed less relevant to libraries, but omitting them is like saying we will only buy indexes and abstracts but no source material, no directories, no reference books. Nonbibliographic database types include directory, numeric, dictionary, and full-text databases.

Statistical and numeric databases present challenges for which many searchers are not well prepared. Edmond Mignon has identified some basic skills which librarians ought to have to deal with these databases, such as: knowledge of content, characteristics, and uses; recognition of needs appropriate to their use; knowledge of how to locate appropriate files; recognition of kinds of needs which require specialist services. He has gone beyond these quite typical categories to list some basic exercises for students of librarianship utilizing online computational procedures in the general reference situation which do not require a specialist's judgment. These include: the retrieval of a specific fact; the retrieval of a collection of facts (an aggregate search); the application of some arithmetical function to the data, such as rank ordering, percent change, per capita conversions, proportions or percentages; and cross tabulation. By advocating these necessary exercises Mignon points out the increasing data-consciousness of the public.[28] These are very minimal skills and seem hardly too much to expect from the coming generation of reference librarians, according to nonlibrarians who use numeric databases. Many academic and research libraries will also want to have at least one resource professional with more advanced statistical skills.

Other Factors Affecting Future Trends

Users will undoubtedly have home terminals and microcomputers as costs of computers and terminals continue to decline and simultaneously as computers become smaller, more sophisticated, and more reliable. It is estimated that by the mid-1980s powerful micros will be available in the $100 range.[29]

At the same time, the growing number of databases (and systems) combined with the continued lack of standardization among database producers will make it harder for occasional users to select databases and plan search strategies.[30] The differences between systems are relatively minor and easily overcome compared to the variety of unique codes-searchable fields and other access points. Within only a dozen databases, Spigai found 93 different fields, 62 of which she thought unlikely to occur in more than one database.[31] This proliferation of searchable codes and elements allows for greater refinement of search strategy but works against use by the end users.

Conclusion

These are rapidly changing times. It has always seemed to me that the way to survive change is to find those higher values which transcend the flux and flow of immediate conditions and circumstances and then anchor your aspirations to those higher values.

Librarians founded library networks because we wanted to make it possible to put any book into any hand. Now, library networks are not dealing with just "a book" anymore, and the network may now be a network encompassing more than libraries. And it may not be free anymore, at least for those who can pay. But I think ALA is unique among the organizations in our field for its commitment to providing information to *all* people. That commitment to free access to information transcends form and source and will be as valid tomorrow as it ever has been. The library, as we know it, has furthered this value, but technological developments will make a far greater store of information available than all but the largest libraries of the past could deliver.

It seems likely that the new information sources will be run on a very rationalized dollars-and-cents basis by businesses whose profit motive will lead them to promote the distribution of information to all those who can pay the price, which will optimize their producers' profits. Our role here is to find ways to provide information to those who cannot afford to buy it. In the past, the means of doing so was the provision of free library service to all. It seems obvious that we will not be able to provide the analogous free information services to all in the future. But the librarian or "public information provider" can have a vital role in obtaining information for those who can't afford their own computer terminals or can't afford the prices of accessing systems. Just as agencies exist to provide medical, legal, and educational services to the disadvantaged, there will certainly be a need for information services and we must explore the means of filling this role. Providing these information services will be perceived as a much more political action, however, than providing free library service to all or even providing free medical service to the disadvantaged. To meet this public need in a nonpolitical, unbiased way will be a tremendous challenge.

Notes

1. Martha E. Williams, "Database and Online Statistics for 1979," *ASIS Bulletin* 7 (December 1980): 89.

2. Martha E. Williams, "Highlights of the Online Database Field–1982," National Online Meeting, *Proceedings* (Medford, N.J.: Learned Information, 1982), p.1.

3. Martha E. Williams, ed., and Sandra H. Rouse, *Computer-Readable Data Bases: A Directory and Data Sourcebook* (Washington, D.C.: American Society for Information Science, 1976); Martha E. Williams, ed., *Computer-Readable Data Bases: A Directory and Data Sourcebook* (Washington, D.C.: American Society for Information Science, 1979); Martha E. Williams, ed., *Computer-Readable Data Bases: A Directory and Data Sourcebook* (White Plains, N.Y.: Knowledge Industry Publications, 1982).

4. Vannevar Bush, "As We May Think," *Atlantic Monthly*, 175 (July 1945): 101–8; J. C. R. Licklider, *Libraries of the Future* (Cambridge, Mass.: M.I.T. Press, 1965), pp. 6–7.

5. Carol H. Fenichel and Thomas R. Hogan, *Online Searching: A Primer* (Marlton, N.J.: Learned Information, 1981), p.2.

6. Charles P. Bourne, "On-Line Systems: History, Technology, and Economics," *Journal of the American Society for Information Science* 31 (May 1980): 155.
7. Ibid.
8. Ibid.
9. Janet Egeland, "The SUNY Biomedical Communication Network: Five Years of Progress and Plans for the Future," unpublished paper (State University of New York, n.d.).
10. Lawrence G. Roberts, "The Evolution of Packet Switching," *Proceedings of the IEEE* 66 (November 1978): 1307.
11. Ibid., pp.1308-9.
12. Ibid., p.1311.
13. Linda C. Smith, "Implications of Artificial Intelligence for End User Use of Online Systems," *Online Review* 4 (December 1980): 388.
14. Information provided by Gregory Benson, BRS, Inc., Latham, N.Y., in an interview with the writer, April 14, 1982.
15. Information provided by Vincent O'Leary, President, State University of New York at Albany, in a letter to alumni and friends, March 12, 1982.
16. Pauline A. Cochrane, " 'Friendly' Catalog Forgives User Errors," *American Libraries* 13 (May 1982): 303-6.
17. Jeff Pemberton, "The Inverted File," *Online* 6 (May 1982): 6-7.
18. Ibid.
19. Mary C. Berger, "The Endangered Species? Can Information Service Survive?" *ASIS Bulletin* 8 (October 1981): 14.
20. Ibid.
21. Paul G. Zurkowski, "The Library Context and the Information Context: Bridging the Theoretical Gap., *Library Journal* 106 (July 1981): 1383.
22. U.S. Congress, House, *A Bill to Amend the Federal Property and Administrative Services Act of 1949 to prohibit Federal agencies from vending telecommunications and automatic data processing services, and for other purposes*, HR 4758, 97th Cong., 1st sess., 1981.
23. "Industry to Feds: Keep Out of Databases," *ASIS Bulletin* 8 (April 1982): 6.
24. Andrew Sherrington, "*Index Medicus* Jeopardized," *Canadian Medical Association Journal* 126 (March 1982): 459-60, as quoted in Albany Medical College of Union University, *Library Newsletter* 10 (April 1982): 3-4; "Industry to Feds," p.6.
25. Sherrington, "*Index Medicus* Jeopardized," p.4.
26. Philip W. Williams and Gerry Goldsmith, "Information Retrieval on Mini- and Microcomputers," *Annual Review of Information Science and Technology* 16 (1981): 95-101.
27. Martha E. Williams and Ted Brandhorst, "Future Trends in A&I Database Publication," *Bulletin of the American Society for Information Science* 5 (February 1979): 28.
28. Edmond Mignon, "Numeric Data Bases in the Professional Librarianship Curriculum: Implications for Behavioral and Social Sciences Librarians," *Behavioral and Social Sciences Librarian* 1 (Spring 1980): 181-7.
29. Marilyn K. Gell, "The Fortune Cookie: Socio-Political Impact of Information Technology," *Special Libraries* 72 (April 1981): 99.
30. Audrey Clayton, "Factors Affecting Future Online Services," *Online Review* 5 (August 1981): 294.
31. Frances G. Spigai, "Impediments to Online Database Use: Size and Its Importance in Database Design and Online Customer Support," National Online Meeting, *Proceedings* (Medford, N.J.: Learned Information, 1982), p.526.

JANE I. THESING

Online Searching in Perspective: Advantages and Limitations

College and Research Libraries News reports that library fallacy twenty-one is "The computer and data bases will solve all information problems." *C&RL News* also asserts, however, that library fallacy twenty-two is "The computer and data bases can solve none of our information problems."[1] Both statements are genuine fallacies, because online searching is not a panacea for all information needs, but it can provide the best technique for answering many of them.

Online search capability has created a whole new range of possibilities for information users. In many cases, what the computer can do seems truly magical. When the searcher has access to a huge file such as the ERIC database, which now includes more than 450,000 items and can retrieve all references relevant to a specific subject in a matter of minutes, the capability of automated information retrieval is indeed awe inspiring.

An online search offers many advantages to users. It is especially helpful for a user with a complex multitopic search. If a user needs to find references on the effect of food additives on the behavior of hyperactive children, he can do it manually by looking through indexes under the indexing terms for food additives, hyperactivity, or children, and scanning the titles for items which might be relevant to his specific interest. He is likely to appreciate that the computer can combine the terms for food additives, hyperactivity, and children, thus instantly matching the three parameters of his topic.

The computer can qualify or limit search results in many ways, depending on individual database specifications. Searches can usually be limited to citations in specific languages and can be qualified by date of publication, and often by document type as well. Thus, a user who wanted to find a review article on a certain topic which had been written in the past two years could qualify the search for these characteristics. The computer can be keyed to a variety of different elements in a citation; the capability to offer access points which are not available in print sources can be valuable for specialized queries. In many databases, such as Social Sciences Citation Index, the searcher can retrieve by name of institution or author affiliation. In some biographical databases, such as American Men and Women of Science, it is possible to retrieve lists such as the total number of graduates of a particular institution.

Another important advantage for an online search is that it can retrieve information on topics which are not included in subject indexes. Many topics are too current to have worked their way into indexing vocabularies. Compliers of database thesauri are not notably faster at updating indexing terms than the Library of Congress. Topics which are currently in vogue, such as burnout, toxic shock syndrome, or displaced homemakers, are often difficult, if not impossible, to find in subject indexes. Specific test names or laws rarely appear in subject indexes. A user who wants information about the reliability and validity of a specific intelligence test would have to search manually through references about all intelligence tests. The computer can pull references to these key names and numbers from titles and abstracts immediately.

An online search service increases the number of sources available to find answers to users' questions. First, online vendors offer access to indexes not owned in print form by a particular library or institution. If a specialized and expensive indexing service has to be discontinued by a library, it is comforting to think that users can still have access to the source online. Second, many online indexes are available only online; they have no print counterpart. Such heavily used databases as ABI/INFORM and PTS PROMPT, business databases, and Exceptional Child Education Resources are in this category.

An online search produces an individualized bibliography designed for a specific query. Libraries which cannot afford to offer manual customized services for their users can appreciate the advantage of having the service available, whether free or for a price.

An online search provides citations that are more current than those found manually. Most database producers update their online files before producing and distributing print indexes. Online citations are cumulated instantly; there is no lengthy month-by-month search of the current year's indexes.

An online search strategy is subject to immediate modification at the terminal. If results are not satisfactory, if a significant parameter of the query has been overlooked, adjustments can be made as soon as the problem is identified. For example, a medical search which retrieves references to studies done on rats, pigs, and chickens can be easily limited to human subjects. A search strategy can be made broader if an insufficient number of references is retrieved; it can be made more specific if too many irrelevant hits are appearing.

Once a search has been prepared successfully, the same search can be automatically repeated against updated files of a database. The capacity to provide this automatic selective dissemination of information (SDI) service is another advantage of an online search. A researcher in chemistry, for example, can receive monthly updates of new titles appearing in his area of interest from Chemical Abstracts database.

By demanding a well-informed search plan at the beginning, an online search forces both librarian and user to think the research need through carefully before plunging into a process that might be vague or misdirected. As librarians participating in the online process, we can appreciate that the thorough analysis of the information problem, which is necessary before going online, often results in a far more rigorous examination of the query than we offer to a user who is planning a manual search. It is temptingly easy to point to the row of indexes and say, Why don't you look in that one? without specifically dealing with either the information query or the nature of the suggested source. An online search cannot be performed successfully without examination and detailed knowledge of both the specific query and the proposed source or database. Moreover, the intellectual exercise involved in setting up a search strategy can bring lasting benefits to users as they confront new questions. Searchers in university libraries have often observed that patrons of search services are much improved in articulating their needs on subsequent information requests, whether using online or printed bibliographic tools.[2]

It is tempting to think of online searching as a magical tool which will effortlessly unlock the mysteries of the universe, but as I pointed out at the beginning of my remarks, the idea that the computer and databases will solve all of our information needs is a fallacy. An online search is not the best answer to all information needs, and online searchers must develop a keen awareness of the limitations of online searching. Indeed, users frequently come to search interviews with gross misconceptions about the very nature of online searching and very unrealistic expectations about what it can provide. These notions must be eliminated firmly, but always diplomatically, in the interchange between user and searcher.

Users often think that an online search can tap one file which contains all knowledge rather than a variety of discrete indexes which cover individual fields well, partially, or not at all. For people with interdisciplinary topics in such fields as women's studies or urban studies, it can be disheartening to realize that database producers are usually tied to traditional discipline divisions.

Users must be aware that the majority of databases available are bibliographic in nature; they produce citations, not facts. Usually they do not retrieve texts of documents (though various vendors offer document delivery services at consistently high prices). It is sometimes even the searcher's lot to dash the hopes of the undergraduate who has postponed his term paper until the last possible moment. No, an online search will not compile information, much less write a term paper!

Another frequently held misconception is the idea that an online search taps only those resources available in the user's particular library or institution, which for most of us is just not true. An online search service produces a list of references which may not be near the user's library or community. Access to the items retrieved may indeed be elusive. It is common for libraries that have started online search services to experience dramatic increases in their users' requests for interlibrary loans.[3]

Yet another misconception is the idea that the computer can pick out the good articles and reject the bad ones. An online search has no quality control capability. A patron who wants "just ten good ones" must be informed that the computer can't tell the difference. Indeed, many users mistakenly invest the computer with magical powers of thought and discrimination. An inarticulate patron with a muddy topic might expect the computer to do his thinking for him. The computer can't and won't. Computers are tools, not creative thinkers, and the results of an online search reflect what is literally requested, not, perhaps, what is wanted. To quote Paul Lacey at the Minneapolis ACRL conference, "Computer databases will be selective and discriminating only if we tell them to be so. They make a craft possible, but they are not a craft in themselves . . . it is a poor craftsman who blames his tools when something goes wrong."[4]

Another limitation of online searching is the fact that most databases have citations from the last ten to fifteen years only. Many are of even more recent origin. Few of them have added retrospective coverage of indexes that were produced before the online era. Thus, I disappointed a prospective user recently whose topic involved finding research done on personality measures before 1959.

Though many fields, particularly the natural and social sciences, are well covered by computerized indexes, and the growth of these indexes is continuing to increase, many fields, especially the arts and humanities, are not nearly as well covered. Film, anthropology, journalism, and literature, for example, are not generally well served by computerized indexes; print sources in those fields are the first necessary bibliographic tools. Another limitation of online databases is their inadequate coverage of monographs. Though journal literature is often covered comprehensively, many databases cover monographs selectively or not at all.

Another limitation of an online search, which often bothers scholars and others used to traditional search methods, is the impossibility of "browsing." The user doesn't get to scan the field of titles which do not fit the search profile. The chances for chance discovery of a new avenue of interest or investigation are thus reduced.

A related drawback is the necessity of having a search intermediary for an online search. Most search systems have rather complex protocols and procedures which must be followed in order to use the system; the search itself is performed by a librarian/searcher who has training and experience in these procedures. Whenever a query must go through an intermediary or middleman, there is a chance of miscommunication and misunderstanding. Of course a thorough search interview can do much to mitigate this problem, but the possibility of misinterpretation persists. There is a chance that some nuance of the query will be lost in translation.

It is important that we, as librarians and online searchers, add our database access to the array of information sources already available; it is equally important to avoid a segmentation which splits the world of the computer from the print world. When a search librarian meets with a prospective search user, the librarian has the responsibility of solving the user's information problem in the most desirable way. Just as a surgeon should not hesitate to recommend alternate treatments which might be more beneficial to a prospective patient, so must an online searcher evaluate other options available to the user before carrying out an online search. An online search is not the best answer for everyone.

The librarian and the user must consider the query in terms of all sources available. There are several relevant considerations. What sources has the user already consulted? If print sources have not been searched, is there one which would easily give an answer? If print sources have been searched, what additional computerized sources are available? What online databases cover the field? What print sources should be consulted in addition to the online sources? Is the user's topic suitable for an online search? Is the topic too broad for an effective search? A user who wants everything ever written about teaching reading or heart disease should be advised that the results would be truly staggering.

Many libraries which charge for online searches have users who are very concerned about what the information will cost. What are likely to be the costs involved in doing an online search as opposed to a manual search? What are the user's priorities? Is comprehensiveness of search a concern? Are monetary costs a concern? Or does the user consider time so valuable that an online search seems inexpensive by comparison? For cost-conscious users, particularly, it is important to point out the availability of appropriate print indexes. Thus, a user who wants to see some articles about the Xerox Corporation may be quite happy to consult *Funk & Scott* or *Business Periodicals Index* instead of the often expensive business databases.

The librarian/searcher has a great opportunity to expand the boundaries of access to information. Information sources, though, should be seen as a continuum, in the words of Simone Klugman, "where no clear demarcation lines can be drawn between computer searching and traditional reference services, and where there is a frequent spill-over of one activity into the other."[5] Both users and librarians will find great benefit from taking this perspective both now and as the options for online searching change and develop in the future.

Notes

1. "More Fallacies of Librarianship," *College & Research Libraries News* 43, no. 4 (April 1982): 125.
2. James A. Cogswell, "On-Line Search Services: Implications for Libraries and Library Users," *College & Research Libraries* 39, no. 4 (July 1978): 276.
3. Ibid., p.279.
4. Paul A. Lacey, "Views of a Luddite," *College & Research Libraries* 43, no. 2 (March 1982): 116.
5. Simone Klugman, "Online Information Retrieval Interface with Traditional Reference Services," *Online Review* 4, no. 3 (1980): 267.

CAROL M. TOBIN

Initial Considerations for an Online Search

The advantages and disadvantages of online searching, discussed above, are the very qualities to keep in mind in deciding when an online search is more appropriate than a manual one. In addition to these somewhat theoretical considerations, there are practical ones relating to staff, patrons, time, and money. Whether or not they are searchers themselves, even whether or not their library offers online services, more and more librarians must learn to evaluate whether or not a request is searchable online.

There are, I believe, seven factors to take into account in deciding if a question is best answered by means of an online search: (1) subject matter, (2) nature of the question, (3) level of need, (4) time, (5) patron preference, (6) completeness of information, and (7) availability and competence of the staff.

> *Subject matter.* If there is no appropriate database for the topic, then all other considerations aside, it is not searchable. There is, for example, no database devoted to classical literature. The trick is to know if there is an appropriate database. This is complicated by the fact that two or three new databases, on the average, become available each month. Yesterday's unsearchable topic may be searchable today.
>
> To learn whether there is a suitable database, you can first use regular reference skills. What index would you use in print; is it available online? There are also guides to databases. *Computer-Readable Data Bases: A Directory and Data Sourcebook* and *Directory of Online Databases* are two that allow a subject approach.[1] In addition, each of the three major vendors offers a quick and economical way to test the number of times a term or search statement appears in a variety of databases. I recently did a search on the Ordos Desert. Since I don't nor-

mally search in the geosciences, I looked for the word Ordos in a number of social science and geoscience databases through use of DIALINDEX. I got no results in some bases and eight and twelve hits in others. I then did the search in the bases in which I had gotten results. With experience, it becomes easier to judge the correct database—sometimes.

> *The nature of the question.* Some questions are particularly appropriate for online searching or can only be answered online. How current is the information needed? The only detailed index to yesterday's *New York Times* is online. On the other hand, most databases do not cover material published before the sixties, so a manual search is indicated if you need articles from the thirties. How complicated is the question? If it is a single-term search of a well-defined topic, for example, biofeedback, it is usually not necessary to go online, but if there are a number of concepts (like the use of biofeedback among women subjects to raise the temperature of the hands) or a number of qualifying parameters like years or language (for example, review articles on biofeedback in German published since 1972) or a number of synonyms to be used (like biofeedback and the aged or senior citizens or older adults), then a search is more efficient. Even a single term can be appropriate for a search if the term is very new and/or not yet well indexed, for example, articles on Rubik's cube. How much trouble would it be to do the search manually? It would be possible to find manually all the U.S. dissertations that mention Philadelphia in their title, but an online search would give the answer in seconds, not days. Is there a print source to use? Some databases do not have a print equivalent, or a given library or branch may not own the print source. Is there any way to get the information needed from the print source? You can verify a citation online when the patron can supply only a few words from the title of an article or manipulate the information in the AMERICAN MEN AND WOMEN OF SCIENCE database to get a list of all the women biochemists in California. Is the information too specific or local? Even though the computer can get to very detailed information, if there is no database that covers your home-town news, then a search won't help.
>
> *The level of need.* Does the patron need an exhaustive search or just five articles? There is a whole database devoted to child abuse, but if a high school student needs to find a few articles to write a five page paper on child abuse, then just opening *Reader's Guide* would

be adequate. On the other hand, a graduate student working on a dissertation might want a comprehensive search of three or four databases and be delighted when nothing turns up directly on the dissertation topic.

Time. How quickly are the results needed? Unless results are printed online, it takes three to five working days for the printout to come. Even with the results of the bibliographic search in hand, there is the problem of document delivery. One can find five dissertations on a subject online in a few minutes. It may take days, weeks, or months to get the dissertations themselves. There is, in addition, the problem of finding the time when searcher and user are both available. Schedules of searcher and patron do not always mesh and a busy search service can get backed up with results. I once had six months elapse between a search request and the search because the faculty member was on leave and rarely on campus.

Patron preference. Some patrons prefer to have an online search done even when it is possible to do the bibliographic search manually, because they feel uncomfortable with library tools, they are pressed for time, they want abstracts to scan for relevance, or they want the material sorted in a certain way. Choice of database may also depend on patron preference, especially in regard to cost. A search that would be better done in the PSYCHOLOGICAL ABSTRACTS database may be done in ERIC if that's all the patron can afford.

Completeness of information. Is the patron too vague? Is the request a search on locus of control? It would turn up a few thousand citations. Sometimes the only way to convince a patron to focus a search is to go online quickly to show the amount of material available. Are you talking to the end user of the search or to a research assistant, secretary, or friend? Search request forms are often helpful in avoiding the problem of incomplete information.

The availability of staff and databases and the competence of the staff. Does the library subscribe to the system with the database that would best fit the search request? If not, a referral should be made. Does the staff have experience in a given system or database? One can be an excellent social science searcher and get lost in a database like CHEMICAL ABSTRACTS. Is the appropriate staff member available when the patron is? In a library with one search analyst not much can happen when that person is on vacation or at a conference.

Spelling this all out may make the decision to search seem more intimidating than it really is. In deciding whether to recommend an online search, you use skills that you already have as a librarian. With experience and the help of database indexes, the question of to search or not to search will not loom so large.

Note

1. *Computer-Readable Data Bases: A Directory and Data Sourcebook* (White Plains, N.Y.: Knowledge Industry Publications, 1982) and *Directory of Online Databases* (Santa Monica, Calif.: Cuadra Associates, 1979–).

CAROL HANSEN FENICHEL

Databases and Database Producers and Vendors

The online industry consists of four basic components: (1) the database producers that create the computer-readable files; (2) the vendors (also called search services) that make the files available on large computers for searching by many people at a time; (3) intermediary organizations such as libraries that employ persons who search the files for information; and (4) end users who request information. The figure illustrates how information flows from the databases through the vendors' computers to intermediaries and ultimately the users.

Most producers of databases distribute them through one of three primary vendors in the United States: Dialog Information Services (DIALOG), System Development Corporation (SDC), or Bibliographic Retrieval Services, Inc. (BRS). Europeans may access DIALOG or SDC using satellite communications or use European vendors, which they call "hosts." Currently, the vast majority of searching is performed by librarians or other trained search analysts working in libraries or information centers.

Notable exceptions to this scheme are the so-called integrated companies, database producers who act as their own vendors. The National Library of Medicine (NLM) and New York Times Information Bank are examples of this type of organization, and there appears to be a trend towards more producers acting as their own vendors.

Databases and Database Producers

Most databases are computer-readable equivalents of printed indexes or abstract publications. Information is keyed once onto magnetic tape or other computer-readable storage devices. The same tape* can be used both to create the online file and to drive a typesetting device

*While magnetic tape is not the only storage medium used, it likely is the most common. To simplify here, then, the word *tape* stands for any storage medium.

Carol H. Fenichel 27

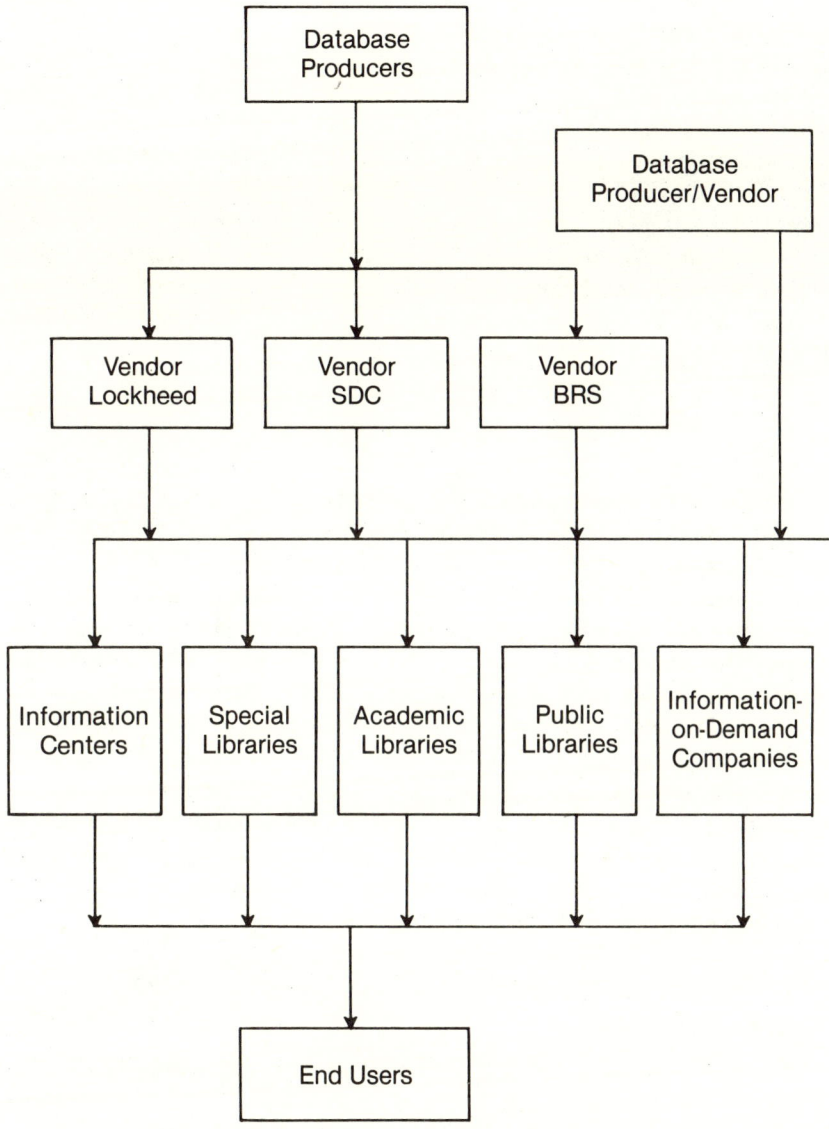

Reprinted from *Online Searching: A Primer* with permission of Learned Information, Inc.

FIGURE 1. Flow of Information from Database Producers

to produce the printed publication. Generally, at the same time a tape is delivered for typesetting, a duplicate is sent to the online vendor to be

28 Technique

used to update the online file. Because it is much faster to update an online database than to publish a printed index, online databases are usually more up-to-date than their printed counterparts.

There are a few databases (e.g., ABI/INFORM) that have no printed equivalent and one—MAGAZINE INDEX—that was first created for online searching and is now available in microform. The EDUCATIONAL RESOURCES INFORMATION CENTER (ERIC) database has the contents of two publications, *Resources in Education* and *Current Index to Journals in Education*, in one file, while some databases are enhanced so that they have more documents online than the comparable print publication. MEDLINE, for example, has a wider coverage than *Index Medicus*.

PARENT ORGANIZATIONS

Producers of databases include not-for-profit organizations, for-profit organizations, and the government. About 40 percent of the databases publicly available on the three major U.S. systems (DIALOG, SDC, and BRS) are produced by not-for-profit organizations, for the most part associated with professional societies. Another 40 percent come from industry; the remainder are supported by the federal government. The scope of the databases coming from all three sectors is varied.

Many databases can be found on more than one system. The MEDLINE database is available on DIALOG and BRS, as well as NLM's own system. NLM and other government-supported distribution systems contain databases produced by for-profit organizations and by societies. Thus, we have a situation in which all three types of organizations are in competition with one another and sometimes an organization of one type even competes with others of the same type.

TYPES OF DATABASES

The first databases searched online in the library setting were bibliographic. Now there are so many types to choose from that it seems that soon it will be feasible to do all reference work online:

Type of Database	*Example*
Bibliographic	ERIC
Directory	Encyclopedia of Associations
Numerical/Statistical	PREDICASTS
Full Text	LEXIS
Book Index	SUPERINDEX

Data	RTECS
Handbook	Gosselin et al.
Chemical substructures	CAS online, DARC
Chemical nomenclature	CHEMLINE
Vocabulary	MeSH
Database Index	DIALINDEX

Other familiar reference works now online are *Ulrich's Periodical Directory* and *American Men and Women of Science*. Several directories of databases are published regularly[1] and the journal *Online Review* carries an updated listing as a regular feature.

SUBJECTS COVERED

Almost all of the library-oriented databases to come online first were in science and technology (ERIC was an important exception). Now it is hard to come up with a subject that is not covered. Major disciplinary areas with examples of major databases in each area are:

Subject Area	Database Examples
Multidisciplinary	CDI, LIBCON
Science	CA Search, BIOSIS
Medicine	MEDLINE, *Excerpta Medica*
Applied Science and Technology	COMPENDEX, NTIS,
Social Science	Sociological Abstracts, PSYCHINFO, LISA
Humanities	Historical Abstracts MLA Bibliography
Business and Economics	Management Contents ABI/INFORM
News and Popular Literature	Magazine Index, NEWSEARCH
Law	LEXIS

Some important indexes from the Wilson Company, such as *Reader's Guide to Periodical Literature*, are not online currently, but are expected soon.

TYPES OF SOURCE DOCUMENTS

Some databases are limited to one type of publication. Others are a mixture of several types. For example, PSYCHINFO covers books, periodicals, technical reports, dissertations, and other monographs. Following are bibliographic databases with examples of source document coverage:

Source Document Type	Example
Books	LIBCON
Conference Papers	Conference Papers Index
Congressional Publications	CIS Index
Dissertations	CDI
Government Publications	GPO Monthly Catalog
Government Reports	NTIS
Journals	Social Scisearch
Newspapers	The Information Bank
Patents	Claims/Citation

MOST POPULAR DATABASES

In 1978, McCarn published what he believed to be the most frequently searched databases in the United States. It is not likely that his list has changed substantially since:

MEDLINE (National Library of Medicine)
ERIC (Educational Resources Information Center)
CA Search (Chemical Abstracts Service)
PSYCHINFO (American Psychological Association)
BIOSIS Previews (Bioscience Information Service of Biological Abstracts)
NTIS (National Technical Information Service)
The Information Bank (The New York Times Company)[2]

It is evident that there is a continued emphasis on science and technology in the online industry.

Vendors

To make databases available for searching, vendors have large computers called time-sharing computers. They can be used simultaneously by many users. The "big three" (DIALOG, SDC, and BRS) acquire databases from many different producers and organize them for searches in the same general way. These systems and the systems of other online vendors have many similarities in the areas of search language, databases available, services, and pricing policies.

MAJOR U.S. VENDORS

Brief descriptions of major U.S. vendors are given below. For a more complete list see the database directories or appendix A of Fenichel and Hogan.[3]

> Bibliographic Retrieval Services, Inc. (BRS)
> 1200 Route 7
> Latham, New York 12110
> (800) 833-4707

Has available over 50 databases including MEDLINE. Connect-hour rates are fairly low if a commitment of 240 hours is made. Without a commitment, a $35 hourly rate is available to which royalties for particular databases are added.

> DIALOG
> 3460 Hillview Avenue
> Palo Alto, California 94304
> (800) 227-1960 (technical help)
> (800) 227-1927 (marketing)
> (800) 928-5838 (California)

The largest of the online vendors with over 160 databases. Also the most widely used vendor.

> System Development Corporation (SDC)
> 2500 Colorado Avenue
> Santa Monica, California 90406
> (800) 421-7229 (Outside California)
> (800) 352-6689 (California)

Over 70 databases on a wide variety of subjects.

> National Library of Medicine (NLM)
> 8600 Rockville Pike
> Bethesda, Maryland 20209
> (800) 638-8480

Contains MEDLINE and 20 other databases in the biomedical area. Costs are lower than commercial systems as they are subsidized by the government.

> Chemical Information System (CIS)
> Information Sciences Corporation
> 2135 Wisconsin Avenue
> Washington, D.C. 20057
> (800) 424-2722

A system with databases containing numeric data is well as bibliographic references. Concentrates in biomedical, toxicological, and environmental areas.

> CAS Online (from Chemical Abstracts Service)
> Online Service/Dept. 88
> P.O. Box 3012
> Columbus, Ohio 43210
> (800) 848-6533

> Questel, Inc. (DARC system)
> 1625 I Street, N.W.
> Suite 818
> Washington, D.C. 20006
> (800) 424-9600

Two very similar and unique systems begun in 1980 for searching chemical substructures. Invaluable when looking for information on classes of structurally related compounds, or for identifying a substance when only the structure is known. These systems are of particular interest to organizations interested in drug design.

> The New York Times Information Service (**NYTIS**)
> Mount Pleasant Office Park
> 1719-A Route 10
> Parsippany, New Jersey 07054
> (800) 631-8056

By far the most comprehensive source of news and general information currently available. Covers the *New York Times,* other major newspapers, and key periodicals.

> Mead Data Central
> 200 Park Avenue
> New York, New York 10017
> (800) 621-0391

Full text searching of LEXIS (a legal database) and NEXIS (a news database). System uses dedicated terminals and is designed to be easy for end users to search.

> Dow Jones News Retrieval
> P.O. Box 300
> Princeton, New Jersey 08540
> (800) 257-5114

News, stock quotes every twenty minutes, historical stock quotes, and other business information.

VENDOR SERVICES

DIALOG, SDC, BRS, and the National Library of Medicine all provide the full complement of services listed below to support their online search activity. Other services provide at least user aids and "help" desk service. These services are either free or inexpensive.

User aids. A good manual that explains how the system works is a must. Other types of user aids are thesauri, classification schemes, word-frequency lists, command charts, database-specific system manuals, source document lists, workbooks, searching tips, and extensive database descriptions.

Newsletters. Usually sent monthly, they are used to keep searchers up-to-date on system activities, features and changes, and to provide information that might help users upgrade their skills. Newsletters are free of charge to subscribers.

Training. The three major vendors offer one- to two-day training sessions around the country on a regular basis. NLM has provisions for training both at NLM in Bethesda, Maryland, and at remote sites.

"Help" desk lines. Toll-free hotline numbers have been listed above with the vendors' addresses. The help desks are always staffed during business hours with pleasant, knowledgeable persons who answer questions about the system and refer business-related problems to the appropriate department. Several of the database producers have "help" desks as well.

SELECTING A VENDOR

Almost every organization that is serious about online searching will want to obtain one of the "big three." The prime consideration is the databases available. While there is considerable overlap among the databases on BRS, DIALOG, and SDC, there are also many that are unique to each system. The vendors will supply lists of databases for evaluation, free of charge.

Cost is another factor. MEDLINE and the NLM databases that do not contain input from other producers are the only clear-cut bargains. The majority of the vendors mentioned above charge on a pay-as-you-go basis and do not have start-up fees or minimum-use requirements. However, many give discounts to organizations that are willing to guarantee a minimum annual use. While up to now, connect charges have been the major costs, charging policies in the last year have been getting more complicated. Some vendors charge more for searches performed offline; page charges for offline searches vary; for some databases there are charges now for printing references online; some connect charges are much lower to subscribers of the corresponding printed index; the new chemical substructure searching systems make charges every time a search of the entire file is performed. These different charging methods make cost comparisons difficult.

Most of the time, subscribing to a system is no more difficult than signing a contract stating that one is willing to pay the bills. Volume discounts can be arranged with vendors directly or through regional consortia, or networks, which offer volume discounts.

One factor that should not make a critical difference in deciding initially between BRS, SDC, DIALOG, or NLM is the system command language, the protocols for "talking" to the system. All the languages are highly sophisticated, work well and are, in general, more alike than different. While aficionados may debate differences, it turns out that most searchers simply become devoted to the language they learn first.

The segmentation of large databases and the amount that can be searched online is a much more important factor in choosing a vendor. The very large databases (such as CA Search, Chemical Abstracts, and MEDLINE) are divided by date into several files that must be searched separately. Vendors make these divisions at different dates, perhaps causing additional expense and inconvenience if searches of them need to cover broad time periods. It is even more of a problem if the earlier portions of the database can be searched offline, as is the case for some files in BRS and NLM. On BRS these offline searches may either be retrieved online the next day or mailed; mailing is the only option on NLM.

Notes

1. *Director of On-Line Information Resources* (Rockville, Md.: CSG Press, updated quarterly); *Directory of Online Databases* (Santa Monica, Calif.: Cuadra Associates, Inc., every six months with quarterly updates); *Eusadic Database Guide* (Oxford and New York: Learned Information, updated annually); *Computer-Readable Databases: A Directory and Data Sourcebook* (Washington, D.C.: American Society for Information Science, annually).

2. David B. McCarn, "Online Systems–Techniques and Services." in *Annual Review of Information Science and Technology*, vol. 13, Martha E. Williams, ed. (White Plains, N.Y.: Knowledge Industry Publications, Inc., 1978), pp.85–124.

3. Carol H. Fenichel and Thomas H. Hogan, *Online Searching: A Primer* (Marlton, N.J.: Learned Information, 1981).

JANET BRUMAN

Physical Requirements: Terminals, Printers, and Furniture

Computer hardware is unfamiliar to most of us and bewildering in its variety. There are hundreds of models of terminals to choose from. I will attempt to outline some points that are targeted specifically at dial-up users of online retrieval systems.

It is impossible to really cover this subject in the space available but these are six areas of consideration: (1) mandatory specifications, (2) configuration and location, (3) printers: thermal or impact, (4) video terminals, (5) transmission speed, (6) modems and couplers, and (7) human factors.

Mandatory Specifications

There are certain basic technical requirements for any terminal that is to be used for accessing the online retrieval systems. If the terminal does not have these specifications, it simply won't work! All of these specifications relate to the communications protocols of the terminal. A glossary at the end of this chapter explains more terms.

- ASCII (American Standard Code for Information Interchange). A way of encoding alphabetic, numeric, and other characters into the binary form used by a computer. All terminals used for the online systems must transmit in 8-bit ASCII code. Most IBM terminals use the EBCDIC code, and cannot be used without a special translator.
- Parity. The regular interspersion of "check digits" in the data as a check for correct data transmission. Either even parity or none (0) parity is required by the online retrieval systems. Odd or none (1) will result in illegible garbage characters being transmitted.
- Asynch. Entry of elements of data is not dependent on being in some

continuous relationship with the main computer. Asynchronous transmission is used by virtually all dial-up systems. Terminals designed for leased-line systems (OCLC, RLIN, LEXIS, and others) utilize synchronous transmission, and cannot be used unless special adaptations and additional modems are obtained, such as with RLIN's RLG-40 model.

Serial. Character mode, serial transmission is required. You cannot transmit more than one bit of information at a time (parallel transmission) over dial-up lines.

Duplex. In full duplex, two-way simultaneous transmission takes place. The online retrieval systems use either half or full duplex transmission. It is advisable to acquire a terminal that can be easily switched back and forth. Avoid modems or acoustic couplers that are half duplex only.

Speed. The vendors currently support only three transmission speeds over dial-up lines: 110 baud, 300 baud, or 1200 baud. These are equivalent to 10 characters per second (cps), 30 cps, and 120 cps, respectively.

Configuration and Location

Every computer terminal or equipment installation has three required component parts: a keyboard on which to input instructions to the host computer; a means of displaying both input and output (which can be either a CRT screen or a printer); and a modem for communication over the telephone lines to the host computer. These three components can be all built into one unit. Examples of this integration include the TI 745 portable printer with acoustic coupler or the new Tymshare Scanset CRT with integral modem. Each component can be separate, such as the Apple microcomputer which has the keyboard and screen as different units. Yet another configuration would be a CRT/keyboard unit, with an attached slave printer and external modem. There are numerous other possibilities. Which configuration you select depends on intended uses, equipment location, portability needs, etc.

When a portable terminal is not required, the purchaser has a far broader selection of models and options. A video terminal can be used; the printer can be either thermal or impact; the acoustic coupler can be a separate unit, or be replaced by a direct-connect modem. Weight or size of components, need for multiple power outlets, or requirements for a specific telephone connection will have to be dealt with only one time for a permanent installation, at the initial installation.

Portable computer terminals have been widely used in libraries for online database searching, frequently for reasons having little to do with portability. The small portable units were for some time the cheapest models available. They were also among the few models available with an acoustic coupler built in, which simplified installation and use of the equipment, and also made the purchase and selection process simpler. Other reasons for choosing a portable terminal are compactness and convenience. They require minimum space and are lightweight enough to be used on any tabletop, so no special stand or table is needed. They can be quickly set up anywhere, and just as quickly removed. If a library is short on space, the terminal can be stored in a closet or cabinet when not in use. Being able to store the terminal conveniently also provides protection from vandalism or theft, if these are of concern.

Of course, there are situations where portable equipment is a must. If searching must be conducted on site in a number of locations, or if it is desirable to do on-site demonstrations for community groups, other campus departments, branch facilities, or professional or user groups, then a portable terminal is the obvious and only choice. There are some drawbacks to the portable models. The available selection of models, options, and configurations is much smaller for portables. Portable units cannot use a video display unit, because video terminals are almost all too large, heavy, and fragile to be safely or conveniently moved. Currently, nearly all the portable terminals utilize thermal printers, which some users dislike.

The search service should be established in a separate room, or at least a partitioned area, off the main reference floor. If you cannot afford partitions, then you probably should not be attempting to set up a new library service as important as this. The room should, however, be fairly close to where reference work is done in your library. Having a room or office with a glass panel or window adds a nice touch. The room should have a door that locks and, if possible, its own light switch. An office with sitting room for two is the minimum space necessary (80 sq. ft.). Room for about twelve is desirable, especially in academic libraries. Storage is needed for search results, forms, publicity materials, and supplies, as well as for manuals, thesauruses, and other published literature used by the service. The following lists the minimum and the desirable furniture for a new service*:

Minimum	*Desirable*
Large desk or table	2 desks or tables (1 for work space)
2 chairs	12 chairs (searcher's chair should have wheels and swivel seat)

*Courtesy of Peter G. Watson

Minimum (cont.)
Small bookcase
Small bulletin board
Wall clock
Output basket or rack (segmented if there is more than 1 searcher)

Desirable (cont.)
2 small (or 1 large) bookcase(s)
2 small (or 1 large) bulletin board(s)
Wall clock
Output basket or rack (segmented if there is more than 1 searcher)

Chalkboard (placed 10 ft. or more away from the terminal)
File cabinet
Microfiche reader
Calculator
Projector and screen
TV monitor or large-screen TV system
For impact printers: automobile vacuum cleaner
For noisy printers: sound-proof canopy
Typist's paper holder

Printers: Thermal or Impact

Every user of the online retrieval system requires a printer as part of the computer terminal installation. It is simply not practical to attempt to use these systems without any means for hard-copy output. The time required to transcribe retrieved data from a screen display is prohibitively expensive, even when only brief information is needed. It is often desirable to retain a copy of the search strategy for the searcher, the patron, or both. It is, moreover, absolutely necessary to be able to document any system malfunction for reporting to the vendor, or equipment malfunction to the service technician.

There are essentially two kinds of printers to consider, distinguished by the way they create the print or image: thermal and impact printers. Both types have their advantages and disadvantages. Some of the factors to consider in determining which is best are: (1) expense, (2) portability, (3) applications, (4) noise, (5) volume of printing, and (6) quality of print. Keeping those factors in mind, let's look at each type of printer in more detail.

The thermal printers use a chemically treated heat-sensitive paper, and create the print image by heating areas of a small silicon wafer mounted on the print head. All thermal printers use a dot-matrix font: the individual characters are created through a pattern of dots arranged in a block (typically 5 × 7 or 7 × 9 dots in each cell). The heat-sensitive paper is disliked by many users, because its slick surface is difficult to write on. It also does not duplicate well on some photocopy machines. The paper deteriorates if stored for long periods of time or at high temperatures; file copies of searches may become brittle or faded after some months. The thermal printers can make only single copies, since there is no way for the print element to "burn" through to multiple sheets of paper. And the paper itself can be expensive to buy, ranging from $2 to $5 per 100-foot roll.

The thermal printers do have some distinct advantages, however, the chief among these being silence. Thermal printers make very little noise at all, and so are ideal for use when they are close to areas where noise would be intrusive (public areas, reference rooms, staff work areas, telephones). Thermal printers have almost no moving parts, and so require very few service calls if proper routine cleaning of the printhead is maintained by the user. Less expensive than comparable-speed impact printers, thermal printers range from $600 to $2000 in price. They are also smaller and lighter in weight; some are as low as 15 pounds, including the built-in modem.

The impact printers work in the same manner as the familiar typewriter: the printhead strikes an inked ribbon, creating the image on the paper. Impact printers can be of the dot-matrix type or can be "character" type. These use a preformed character, just as a typewriter does, to strike the ribbon. The print element of a character printer can be in the shape of a golf ball, a thimble, or a "daisy wheel."

The character-impact printers have excellent print quality, comparable to the best office typewriters. These letter-quality printers are only rarely used in online searching applications, because they are expensive ($1800 to $5000) and low speed. Most character-impact printers have speeds ranging from 10 cps to 50 cps. This is generally too slow to be practical for online searching; if the telecommunications network is transmitting data continuously at 30 cps, the printer's speed must exceed this to keep up. This is particularly true if output will contain many blank lines, which slows down the printer throughout even more. Character-impact printers are used more for word processing applications than for database searching.

Dot-matrix impact printers are the most widely used in libraries, as well as the data processing industry as a whole. They are fast, flexible, and moderately priced. The dot-matrix impact printers fall into a middle price range of $1500 to $3000, more than most thermal printers but less than the character-impact printers. Print speeds range from 10 cps to 180 cps. Most use any standard sprocket-feed paper, either single sheet or multiple copy. Dot-matrix impact printers are durable and will have a longer life span overall than the thermal printers. They come with a wide variety of options, including adjustable carriage width, forms feed control, variable print sizes and fonts, special or expanded character sets, etc. Impact printers create a sharp, clear image that duplicates well, although not everyone finds the dot-matrix fonts attractive. Newer models are introducing "double-pass" printing, which enhances the attractiveness and legibility of the type but slows down throughput considerably. Some dot-matrix fonts lack descenders on lowercase characters, so that the printout is hard to read.

Impact printers are usually larger than the thermal printers and much heavier (40 to 50 lbs.). Consequently, once in place they are not readily moved about. At present, very few impact printers are available with an acoustic coupler or modem built in, and so the addition of this as a separate item also restricts the movability of the printer. Probably the most serious drawback to the impact printers is their noise level. Continuous printing will result in a noise output of around sixty decibels. Sound-deadening pads or enclosures will reduce the level, but the printer should still be somewhat isolated from public areas, telephones, and other work areas. The best installation would be in a separate room or partitioned area.

Video Terminals

While the video terminals or CRT displays are not required for access to the online retrieval systems, they are often used. Video terminals can *not* be used effectively by themselves and are usually attached to a receive-only by-printer. For those libraries whose equipment budgets will allow the addition of another piece of hardware, a CRT can be a considerable advantage. Terminal operators will find that they much prefer viewing searches on a screen to trying to peer around the printhead of a printer, which too often obscures the characters last printed. When corrections are made to input on a CRT, the corrected characters will be legible, which they are not when an operator backspaces and overstrikes them on a printer terminal. A CRT allows more observers to see what is happening in the search and can be used for group presentations. Demonstrations for more than four people are quite awkward with a printing terminal alone.

Video terminals are especially desirable if other applications are planned for the equipment. Word processing, for example, is nearly impossible without one. Video terminals are much more flexible than printers. They are available with a broad array of features ranging from text editing and offline memory to programmable function keys and graphics character sets. Nearly all operate at 9600 baud or above. A CRT need not be an expensive addition to an online search station. A typical "dumb" CRT can be priced as low as $600. The intelligent models will run up to $1500 or more; libraries will want to investigate carefully whether their money is better placed in a "smart" CRT or in a "smart" modem. The decision will depend on whether the principle use of the equipment is for in-house functions (data entry, word processing) or for accessing dial-up information systems where communications functions become more important. One word of warning: the kind of printer port on different CRTs is not standardized, and the CRT you want may not be compatible with the printer you like best.

Transmission Speed

You have probably all heard a lot of talk about 300-baud versus 1200-baud equipment. For many years, the standard equipment was 300 baud because it was the best phone lines could provide for dial-up service. Then around 1978, 1200-baud modems and phone lines became available and there was instant controversy: which was better for online searching? Because it is four times faster, 1200 baud would seem the obvious choice to reduce connect-hour charges. But the equipment itself (both the modems and the faster printers to keep up with them) cost from twice to three times as much as 300-baud equipment. There was also a tendency to overlook the fact that not all the connect time was used transmitting data; the largest percentage of connect time is generally used by the searcher to input strategy or to contemplate the next move. Several cost-comparison studies concluded that economically, 1200 baud makes sense only in situations where more than forty searches a month are done or there is a very substantial amount of online printing of results.

However, there are now some additional factors that must be added to these comparisons. The search-system vendors and database producers have reacted to the prospect of searchers retrieving information in shorter connect periods by instituting new charging formulas. There are now per-citation "online type" royalties on about one-third of the bibliographic databases. These charges have eliminated most incentives to do all printing online. Some vendors are also charging substantially higher connect-hour fees for 1200 baud than they do for 300-baud access. So far, these speed-dependent pricing algorithms have not been adopted by the major library vendors such as DIALOG, BRS, ORBIT, or NYTIS, but there is no assurance that they will not be in the future.

The development of "intelligent" equipment has also affected the speed of search input. There are now numerous options in intelligent terminals for the online searcher to prepare strategy offline and give it to the search service at maximum speed. This process reduces substantially the amount of time spent online. Even at 300 baud, the savings can be impressive. After all, how many searchers can type at 30 characters per second? That is about 300 words per minute.

Both of these new considerations must now be included in deciding what speed equipment to buy, along with the old consideration of price differential, which still holds true: high-speed modems and printers are roughly twice as expensive as low-speed ones.

Modems and Couplers

Perhaps the most rapid technological changes in the terminal market have been occurring in the area of the communications interface. Not so long ago, the available choices were limited to a low-speed acoustic coupler (built in, if the terminal was a portable model) or a separate modem with a special telephone. The acoustic couplers, because of their "ear muffs," could be used with any telephone but provided little data integrity. The separate modems were less susceptible to line noise, because they were wired directly to the telephone line. The special "data phones" or "exclusion-key" phones (which had voice-to-data switches built in) were expensive, however. High-speed transmission was available only with the separate modems, not with the acoustic couplers, and it was necessary to choose between one of the two communications protocols, either Bell 212 or Vadic 3400 series. The local Tymnet or Telenet node usually determined the choice of speed and protocol.

The available selection is now much more varied than it was just a few years ago. Acoustic couplers are now available in both 300 and 1200 baud. Separate modems no longer require exclusion-key telephones; instead they are "direct connect." And you no longer have to choose between the two protocols: modems are available that are compatible with both Bell 212 and Vadic 3400 series. The network nodes also have dual compatibility.

External modems, not built into the terminal, remain standard but few require the use of an exclusion-key telephone. These modems are now being designed with direct-connect modular telephone connectors, which allow them to be plugged into any standard RJ11C telephone outlet. (This is the same small plug-in jack you will find on all newer residential and single-line business installations.) The old voice-to-data switch is now built into or attached to the modem rather than the telephone. The direct-connect jacks allow the user to install and move the modem, since no specially installed exclusion-key phone is required, and also allow the user to avoid the higher monthly line charges associated with the data phones. Generally, the choice now will be between an acoustic coupler or a direct-connect modem. The acoustic couplers are more portable, since they can be used with any telephone. The direct-connect modems provide a connection less susceptible to line noise, but cannot be used with older telephone installations or multiple-line telephones (except with a special adaptor). Many terminals, especially the thermal printers and video display units, are now available with built-in direct-connect modems. There are advantages to the integral or built-in modems and couplers. They are more convenient if the terminal is to be mobile at all. They require no extra space and eliminate some of the clutter of cables and power cords in a permanent installation. They simplify the selection and purchase process by eliminating one component and are also covered by the same maintenance agreement that covers the terminal.

There are, however, advantages to separate units. Many more configurations are possible if the modem is not integral, or built in. Having a separate modem allows the device to be "shared" or switched between more than one terminal. The separate units offer a wider choice of both modems and terminals to choose from, and thus more competitive prices. The separate modems are more apt to offer features such as automatic speed selection, automatic dialing of frequently called telephone numbers, redialing of numbers that are busy or don't answer, or routing a call to an alternate number if the first is busy or doesn't answer. Intelligent modems, having all of these features plus memory and text editing capabilities, are another recent development. They can reduce on-line charges by providing offline input and error correction of search strategies or text.

Another feature to consider is whether the modem should be "originate only," or should have auto-answer. Auto-answer allows incoming calls to be received without an operator attending the terminal. With the growing use of TWX-compatible terminals and modems for interlibrary loan and with electronic mail replacing existing TWX equipment, these capabilities are quite useful for a library.

Practical Design Features

A number of other practical design features should be considered when a terminal is being selected, as operator convenience can contribute significantly to satisfaction with the device.

For both video display and printing terminals, look at the layout of the keyboard. The location of symbols (slash, equal sign, "at" sign) and of special keys (break, escape, control, carriage return) are not standardized, and some keyboards are amazingly ill designed (e.g., having "break" where the "shift" key would normally be). Check the location of the configuration switches. Are they on the rear of the terminal, on the front, or under the hood? Or is configuration done from the keyboard? Can the configurations be changed while the terminal is online? This is not necessary but is definitely nice.

For video terminals, is the keyboard attached or separate? Is the screen adjustable to protect against glare? Does it tilt or have a protective filter? Are brightness and contrast controlled by the operator? How clear are the characters? Any video terminal used for the online systems should have at least an 80-column by 24-line display to avoid wraparound in which the second part of an output line is displayed below the first part and to allow an adequate amount of data to display. This size is standard for most terminals, but not for microcomputers or word processing systems.

For printer terminals, check the operating noise level, preferably at your own location. How complicated is it to load the paper? How easy is it to read while the terminal is printing? Does the printhead or terminal cover obscure your view? For impact terminals that use fan-fold paper, is there a paper catch tray? Are supplies for the printer (paper, ribbons) readily available? For 1200-baud printers, make certain there is a "print" or "receive" buffer large enough to prevent data loss; 2000 characters or more is desirable.

All of these aspects of terminal design should be checked, if possible, by the staff who will actually be using the device.

If you are selecting several components (i.e., CRT with slave printer and external modem), make certain they will be compatible with each other, either by demanding a no-obligation trial period or by including a compatibility clause in your purchase order. And finally, don't buy anything for which local service is not available!

Acquiring the Terminal: Approaches and Sources

Now that we have discussed what equipment you want to acquire, the final question is how to go about the acquisition process. The appropriate method for acquiring your hardware incorporates several factors including the amount of money available, internal accounting or budgeting policies, tax status of your institution, long-range automation plans, etc. These factors will have more significance for some institutions than others, but everyone should at least consider them.

LEASING

Leasing arrangements offer a limited variety of equipment, but shift some of the burden of selecting the configuration from your staff onto the vendor. Leasing may allow the library to obtain a far more expensive and sophisticated configuration, because the total amount does not need to be budgeted in one fiscal year. For some institutions, ongoing costs are easier to budget for than one-time capital expenses; for others exactly the reverse is true. Leases can have varying time periods, ranging from one year to three years or more. Most contain a cancellation clause and/or an exchange clause, which allow you to return the equipment and replace it with something else. Given the rapid technological development in the industry and the expanding applications for automation of library operations, this option can make leasing quite attractive. Some leasing arrangements allow for a purchase option, with some percentage of the monthly lease payments being credited against the eventual purchase. These clauses are not standardized so be extremely careful to read the fine print! If your organization has a legal office, have a staff member go over the contract with you.

RENTING

Renting computer equipment is probably the most expensive method of acquiring a terminal and probably should not be considered by most institutions, except in special circumstances. Short-term rentals (i.e., month-to-month rentals or under-twelve-months leases) trade rather high price tags for user convenience. There is very little commitment entailed; if you don't like the equipment you can get rid of it quickly. If you only need the equipment for a temporary installation, while waiting for capital funds to be released, for a one-time project, while waiting for purchased equipment to be delivered, rentals can fill your needs. The variety of terminals that will be available for short-term rental may be quite restricted, depending on where you are located. Rental agreements virtually always provide for service or replacement of inoperative equipment.

PURCHASE

In the majority of cases, the decision will be made to purchase the equipment outright. Purchase is generally the most economic acquisition method in terms of total cash outlay. Purchasing the terminal offers the widest possible selection of equipment and therefore the greatest flexibility for configuration. The amount of staff time invested in the selection process will be greater initially, but once it is done, it's done. Of course, purchasing your equipment does lock you into a configuration, for at least some time. Because of the rapid technological advances in the industry, there is only a limited market for used equipment. You should not plan on financing a new installation by selling an old one; instead, consider your needs for upgrading before you purchase and select a model with that in mind. One caveat about purchasing equipment: unlike leases or rentals, the responsibility for maintenance is 100 percent the owner's. Investigate the availability and cost of service in your local area before committing yourself to any specific piece of hardware or to its supplier.

This raises another aspect of the terminal acquisition process—the choice of a supplier. Computer terminals can be acquired directly through the manufacturer, through authorized distributors, through various types of service organizations, or through mail-order houses. Shoppers in large urban centers will probably be able to deal locally with any of these; others will have fewer choices. Chances are that none of the suppliers will know very much about the specific library operations or systems you will be using the terminal for, and so you will need to be well informed yourself before talking to them.

MANUFACTURERS

Manufacturers typically will not offer any price reductions on single-unit retail sales; you can expect to pay full list price. However, the manufacturer, who knows the product better than anyone else, has a vested interest in satisfying purchasers. Manufacturers generally offer good after-sales support. Any one manufacturer will probably not be able to supply all the components (CRT, modem, printer) for your configuration, so the burden of determining compatibility will be on you. Some manufacturers handle leases or rentals, but many do not.

DISTRIBUTORS

Authorized distributors and the service organizations almost always carry several manufacturers' product lines and so are in a better position to put together a package to fit your needs. Many of them will offer reduced prices, even on single-unit sales. Since they carry so many models, there is the chance that they will not be thoroughly knowledgeable about the operation or capabilities of every item, although this will vary. Some people feel that because many manufacturers' products are carried, the distributor will be less biased toward one or another. This may or may not be true; you don't know what inventory problems or sales incentives the dealer may have. The real advantage of the distributor's wide product line is that he will be able to determine the compatibility of the components for you and, in fact, has a strong vested interest in doing so. Both distributors and service organizations also frequently handle their own maintenance operations, and so you can expect these dealers to give after-sales support, for a price. Where service is not handled locally, after-sales support will be more varied. Leasing and renting are handled by both distributors and service organizations.

MAIL ORDER

The mail-order houses will usually offer the most competitive pricing. They deal exclusively in sales—no leases or rentals. Their advertisements can be found in journals such as *Byte, Mini-Micro Systems, Interface Age,* etc. While the prices are attractive, it is entirely up to you to know what you want. This means determining specifications, options, configurations, and so on. There is very little likelihood that the mail-order supplier will consult with you in any way, either before or after the sale, but if you are confident enough in your own decisions, you might consider using a mail-order house. Maintenance information will be up to you; make certain you also check on the shipping and warranty coverage for the equipment.

SERVICE

Service contracts should be regarded as an essential part of the terminal acquisition process. With a service contract, you are assured that some level of support will be available to you, on demand, usually within forty-eight hours of placing the service call. Without such a contract, you have no such assurance. Instead, you will have to locate a service outlet each time you need one and then wait your turn. You may have to pack the malfunctioning unit off to the service center, rather than having it repaired on site. The turnaround time may be a week or more, during which time you will be out of operation.

Service contracts will cost between $20 and $50 per month, per item; if you have a CRT and a printer, you will pay for each piece. Make sure you have both covered by the same agency, so there can be no "finger-pointing." You usually will have the option of on-site or depot service. If you are located particularly near the service depot, you might want to consider the savings of delivering and picking up the unit yourself. Doing so will lengthen the turnaround time in most cases. Service contracts should be initiated at the time the equipment is purchased. If you wait until after the warranty has expired, you will have to pay for an on-site inspection (about $300) before any contract will be signed.

48 Technique

Glossary

Acoustic Coupler. A device which transforms the terminal's electrical signals into audible signals that can be carried by the telephone line, and vice versa. See also Modem.

ASCII. American Standard Code for Information Interchange

ASR (Automatic Send Receive). A terminal that can transmit a previously prepared message, from a tape or other storage or memory device.

Asynchronous Transmission. Also called Start-Stop Transmission. Where each character is preceded by a "start" signal and followed by a "stop" signal.

Baud Rate. The speed at which the electronic signal is transmitted from terminal to computer, expressed in bits per second.

Buffer. Storage space used to compensate for differences in rates of transmission between one device and another.

CPS. Characters Per Second.

CRT (Cathode Ray Tube). A terminal with a video screen.

Data Set or Data Phone. See Modem.

Dot Matrix. A type of printer, which forms the characters or letters by using a pattern of dots. Thermal printers usually are dot matrix.

Duplex. See Full Duplex; Half Duplex.

EBCDIC. Extended Binary Coded Decimal Interchange Code.

Full Duplex. A technique of operating a communications circuit (between terminal and computer) whereby each end can simultaneously transmit and receive. See also Half Duplex.

Half Duplex. A technique of operating a communications circuit (between terminal and computer) capable of transmitting and receiving but only in one direction at a time; in contrast to Full Duplex.

KSR (Keyboard Send Receive). A terminal that can both transmit and receive.

Modem. It converts the terminal's digital signals into analog signals that can be carried by the telephone lines, and vice versa. Term is a contraction of "modulator/demodulator." See also Acoustic Coupler.

Parity Check. Addition of noninformation bits to data, making the number of bits in a grouping always odd or always even.

RO (Receive Only). A terminal (usually a printer) without a keyboard. See KSR.

RS232. The designation for the standard port required for connecting the terminal to the Modem.

Slave Printer. An RO printer attached to a CRT terminal.

Synchronous Operation. Where the computer's basic operation is constrained to start on signals from a clock and must keep in step with these signals.

VDU (Video Display Unit). Any terminal having a screen, like a CRT or television.

LAWRENCE R. MAXTED

The Interview Process in Online Searching

The online database searcher is an analyst who, rather than merely typing or translating for the end user, analyzes the request and then restructures it as a search.[1] This role requires that the searcher have, in addition to a thorough knowledge of online searching, the ability to draw from the user a detailed, accurate, and comprehensive description of the need and then to negotiate an online search strategy which matches that need. The searcher's success in accomplishing these tasks depends on the interview.

The intent of this paper is to describe in detail the interview process that takes place during online database searching and to define its purpose. The interview process can be divided into four distinct segments: the initial reference encounter, the presearch interview, the online search itself, and the postsearch evaluation and use of search results.

The Initial Reference Encounter

The initial reference request normally takes place at the reference desk with the user either specifically requesting an online search or asking a question that indicates to the librarian a possible need for an online search. This is the time when online searching can be briefly explained in relation to other methods of gathering information on the request. It is also the time to begin preparing for a search.

To set online searching in its proper perspective to other information-gathering methods, it is necessary to explain what online searching can do that other bibliographic search methods cannot do. This includes contrasting the unique benefits of online searching (the ability to search on keywords, combine major subject terms, and save time otherwise spent in searching manually) against its drawbacks (limitation of coverage, the possible retrieval of irrelevant items, delays in scheduling a search and in receiving the results if printed offline, and any fees). In many instances, it is done in order for the librarian to recommend an alternative to online searching. The user, too, may decide after learning about online searching that a manual search would be more appropriate. In other instances, a search may seem appropriate, but the librarian may recommend that it be postponed because the user is not sufficiently informed on the topic to formulate a search request properly.

If an online search is decided upon, it is then necessary for the librarian to elicit from the user a detailed search request. The request should include a statement of both the specific topic of the search and the required level of the search—whether the search is intended to be comprehensive research or simply to identify a few citations on a narrow subject. The librarian should also request the user to specify as many keywords as the user can think of for the topic. Such information will permit the person who actually conducts the search to research the question and prepare a preliminary search strategy. The initial reference request phase of the online search concludes with making an appointment for the search.

The librarian handling the initial request may or may not actually conduct the online search, but it is necessary for this person to have sufficient knowledge to explain online searching and to determine if an online search is necessary, if such advice is needed immediately.

The Presearch Interview

The presearch interview immediately precedes the online search itself. It provides an opportunity for the searcher and the user to discuss extensively the user's online request and online searching in general. A large portion of the presearch interview, and for that matter the entire online search interview, is instructional in nature. A joint study during the early 1970s by the University of California at Los Angeles and the University of Georgia found that a substantial part of the searcher's role was to instruct or advise users as to database vocabulary, subject coverage, indexing conventions, and data content.[2]

One of the first steps for the searcher in the presearch interview is to have the user restate the information need and explain any concepts with which the searcher is unfamiliar. In turn, the searcher should "echo" the request to the user for clarification and continue doing so until they agree that the searcher understands the request. This restatement of the search topic gives the searcher a clearer idea of the request and insures that the searcher did not misinterpret the user. This interchange between searcher and user also provides a beginning for the dialogue that will continue throughout the search.

At this stage it is necessary to ask whether the user understands the fundamentals of online searching, a question which should have been touched on briefly during the initial reference encounter. If not, the searcher should explain the mechanics of online searching, its unique benefits, and the costs in comparison to other information-gathering techniques. Enough time should be allowed for the user to ask questions about online searching and to personally clarify how the process in general works. The intent of this explanation is to give the user enough background to make wise decisions when negotiating a search strategy with the searcher during the online session. The searcher should also, early in the presearch interview, assure the user that the user is in control of the search and may therefore stop or redirect the search if that becomes necessary.

The most important part of the presearch interview is, of course, the preparation of a search strategy. The user's needs and possible preconceived expectations must be assessed against what the searcher knows it is possible to accomplish through online searching. User misconceptions about online searching may be present in the search request. The user may request a search that is very narrow in scope while really wishing to retrieve all citations even remotely related to the topic; another user may, on the other hand, request a search very broad in scope while wanting to retrieve only the most relevant citations.[3]

The searcher would normally have spent some time before the presearch interview reviewing and selecting databases for inclusion in the search as well as listing suitable keywords and controlled subject terms. Each database presented should be explained in relation to its subject coverage, the type of documents covered, the time period covered by the documents included (how current is the coverage), and the costs, if fees are charged the user. The searcher should also inform the user if there are paper equivalents of the databases being considered that could be consulted. In addition, the user should be told whether the citations contained in the databases pertain to documents that are likely to be locally accessible or not. These are all usually important considerations for the users.

The differences between controlled vocabulary and keywords from titles and abstracts should be explained in order to select index terms. For controlled index terms, the user should be shown any scope notes available from thesauri or examples of documents indexed under the selected term in a database's paper equivalent.[4] The searcher and user should also discuss the specificity of any keywords being considered. While it may seem obvious to a user that a search using the keywords *women* and *draft* would retrieve citations pertaining to women in the armed forces, the search might also retrieve citations referring to the draft of the Equal Rights Amendment for women. This discussion of controlled index terms and keywords permits the user to exclude from the search those terms that are inappropriate and to bring up or, possibly, discover new terms.

Once general agreement has been reached on the index terms and keywords to be included in the search, it is necessary to formulate the search strategy. Some searches may be straightforward retrievals of unique terms, such as a search for all citations dealing with a particular person, thing, place, or event. Searches for all citations pertaining to Richard Nixon, spittoons, the Falkland Islands, and Conway's Cabal are all examples. Such searches could be conducted using one retrieval term or a combination of several synonyms, and they are therefore fairly easy for a user to understand.

Users generally have a harder time trying to conceptualize more complex searches which require the combination of several terms. A search for citations dealing with women and the draft, for example, requires combining, through Boolean logic, terms retrieving citations on women with those pertaining to the military draft. Venn diagrams represent an effective means of both explaining Boolean logic and of describing a search strategy to a user.[5] Figure 1 is the Venn diagram showing the strategy for retrieving articles dealing with women and the draft. The Venn diagrams used during the presearch interview to negotiate the search strategy have continuing value during the search itself, for both the searcher and the user, to keep track of how the search is proceeding.

During the presearch interview, the online search query is trans-

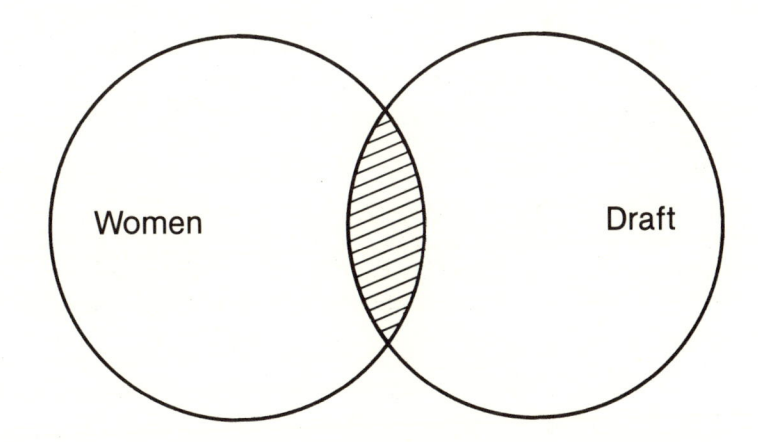

FIGURE 1. Venn diagram for search on women and the draft

formed from the user's initial natural language request, through negotiation, to a series of controlled index terms and keywords which can then be combined with Boolean logic to conduct the search.[6] The results of the presearch interview are a well-formulated search strategy and a sense of cooperation between the searcher and the user.

The Online Search

If the user is present, the interview process continues during the search. This is a crucial phase of the interview process, because it permits an online search to be truly interactive. The user is able to evaluate the results of a search as it progresses and is thus able either to continue a successful search strategy or to alter it when unanticipated problems arise. The user needs to be briefed on the mechanics of the search at the outset and kept informed of the search's progress at each turn.

Before turning on the equipment and conducting the search, it is necessary for the searcher to instruct the user in the mechanics of the online search. This instruction includes familiarizing the user with what may be strange equipment and explaining the basics of the search language being used. A written search strategy, such as the one suggested in the form of Venn diagrams, would also facilitate the user's ability to follow the search and make decisions along the way.

As the search continues, the searcher should keep up a dialogue with the user as to what is happening and what is being retrieved or not being retrieved. The user may be able to suggest ways to widen or narrow the search. It may be necessary in some instances for the searcher to suggest stopping the search and re-evaluating the search strategy. A temporary halt may also be warranted when time is required for the user to evaluate the results and, for example, to decide whether to have a group of citations printed online or offline.

At all times during the search, the user needs to understand what is taking place and should be in charge of the search while the searcher provides the technical work necessary to carry out the user's commands and to advise the user on those commands and the search strategy.

The final stage of the online portion of the interview comes when the search is terminated because of either success or failure.

The Post-search Interview

The purpose of the post-search interview is to evaluate the results of the online search and to provide the user with a bridge from the online search to the gathering of the documents discovered. The bridge usually includes instruction in information-gathering methods.

The postsearch interview may take place immediately following an online search or at a later date, when offline printed citations are picked up by the user. Interviews could also take place at both settings. The primary intent of the postsearch interview is for the searcher and the user to evaluate the citations retrieved by the search. One important consideration in evaluating the citations retrieved is to determine how relevant they are to the topic searched and, therefore, how precise the search was. The user may also be concerned about the quantity of citations retrieved, depending on the project requirements.

The postsearch interview is also an occasion for the searcher to explain the search strategy again and to indicate why certain types of citations were retrieved while others were not. For example, a search on women and the draft limited to current-affairs databases would not retrieve as many historical articles as would a search that included the historical databases. To fill these research gaps, the searcher might then outline strategies for possible future searchers.

Since many online searches result in the retrieval of large numbers of citations, many of which may not be held locally, the searcher needs to provide the user with a start in tracking down the actual documents cited. The first step is for the searcher to go over the citations with the user and explain abbreviations when necessary. A brief introduction to the various tools for retrieving documents in the library and through interlibrary loan may be needed. This phase of the postsearch interview should assist the user back into the mainstream of the library's document retrieval system with the least amount of confusion to the user and to the system.

The searcher should ask the user during the postsearch interview to report any large number of documents, relevant to the topic and found later, that were not indicated by the online search.[7] This permits an analysis of why the search missed the documents in order possibly to reformulate the search. It also builds the searcher's understanding of the databases in question.

Conclusion

Although this paper breaks down the online search interview into four ascending stages, these phases may not be as clearly delineated in practice. Some stages may be touched on only briefly during a search. For example, a user may only require a quick search of a database for a specific citation. In such an instance, the four stages of the interview would be greatly compressed and hardly distinct.

The interview process should be flexible enough to adapt to different users' needs and expectations. The constant task is to determine accurately the user's need from the request and then to satisfy that need via the online search. To satisfy the request requires the linking of the online search skills of the searcher with the user's own subject expertise via the dialogue created and maintained during the interview process.

Notes

1. Sara D. Knapp, "The Reference Interview in the Computer-Based Setting," *RQ* 17 (Summer 1978): 321.

2. W. C. Zipperer, *User Interface Models for Multidisciplinary Bibliographic Information Dissemination Centers* (Bethesda, Md.: ERIC Document Reproduction Service, ED 122 846, 1975), p. 14.

3. Janet M. Dommer and Dawn M. McCaghy, "Techniques for Conducting Effective Search Interviews with Thesis and Dissertation Candidates," *Online* 6 (March 1982): 45.

4. Ibid., p. 47.

5. Sallye Wrye Smith, "Venn Diagramming for On-Line Searching," *Special Libraries* 67 (November 1976): 511–15.

6. Susan Artandi, *An Introduction to Computers in Information Science*, 2nd ed. (Metuchen, N.J.: Scarecrow Press, 1972), pp. 69–70.

7. Arleen N. Somerville, "The Place of the Reference Interview in Computer Searching: The Academic Setting," *Online* 1 (October 1977): 20.

KRISTINE SALOMON

The Mechanics of Online Searching

Online database searching is becoming a valuable asset to libraries, businesses, and universities. This paper will describe database selection and search strategy for the novice searcher.

A successful search relies on many different factors, the most critical being total comprehension of the search request. The search analyst must understand what the user is requesting in order to develop a search strategy that will produce satisfactory results. If the searcher is unfamiliar with the topic, three options are available: (1) consult an expert in the field to obtain background information, (2) do background reading on the topic, preferably recommended by the patron requesting the search, and (3) refer the search to someone who is more familiar with the topic. Complete comprehension of the request is critical for satisfactory results.

Database Selection

With over 150 databases to choose from, the searcher may feel overwhelmed by the possibility of selecting an adequate database for the given search. However, Fenichel and Hogan have outlined a few guidelines for database evaluation:[1]

> *Subject coverage.* Many times the selection of the database is obvious by its subject coverage. For example, if one is doing a search on management styles, ABI/INFORM and Management Contents would be clear choices. However, if the patron is interested in finding material on television in the Soviet Union, many databases may be useful, such as PAIS, GPO Monthly Catalog, Social Science Citation Index, or Dissertation Abstracts.

The three major vendors have directories of databases online which may assist the searcher in the selection of appropriate databases. DIAL-

INDEX from DIALOG, CROS from BRS, and Data Base Index from SDC each identifies databases in which a term is used and indicates the frequency of that term in the database. A study of these directories is an excellent way to select a database when there is no obvious first choice.

Type of source document. Books, journal articles, technical reports, newspaper articles, and dissertations are a few of the many different types of items retrieved from a search. Some databases include only one type of document, such as Dissertation Abstracts, while others retrieve different types of documents. It is important to know what the patron is interested in receiving.

Availability of source documents. With millions of bibliographic citations online, and in excess of 55 million records on DIALOG alone, some material retrieved from the search may not be readily available. Libraries with specialized collections in business or education, for example, are not likely to subscribe to all of the journals indexed in databases that are relevant to those disciplines. Although a library with a specialization in education may have a complete set of ERIC documents on microfiche, the possibility of that library having access to the more than 700 periodicals indexed in ERIC is more remote. Two databases that address management interests, ABI/INFORM and Management Contents, scan over 550 primary publications in business and related fields, and over 400 domestic and international journals, proceedings, and transactions, respectively. Availability of source documents is even more tenuous in topics of an interdisciplinary nature that demand that an online search be conducted on multiple databases in several subject areas. Interlibrary loan services and vendor ordering systems, such as DIALORDER, facilitate document retrieval. However, if the patron needs the documents immediately, the searcher may choose a database that will retrieve citations more likely to be found nearby.

Cost. Database charges are computed by cost per connect hour and thus vary considerably. A search can cost anywhere from less than one dollar to hundreds of dollars, depending on database connect cost, royalty fees, time spent on the computer, and amount of information retrieved.

Coverage dates. Some databases retrieve material written in the 1800s while others may retrieve items only from the past few months. The nature of the search will determine the appropriate database.

Currency. A database may be updated daily, monthly, or annually. The frequency with which the database is updated is an important consideration to remember if the patron is requesting information on a current topic.

The three major vendors have produced guide sheets on each database which indicate the types of documents retrieved from the database, the dates of the documents, and the updating policy. The guide sheets also describe the search capacities of the database. They are essential for database selection.

Another important consideration in database selection is whether or not the database has a printed equivalent. The searcher and user may be able to determine if appropriate material will be found for the given topic by examining the print version before going online.

The searcher may wish to select more than one database for a particular search. Multiple database searching has the advantage of potentially retrieving more relevant material, but it has the disadvantage of duplication. However, a higher degree of user satisfaction has been reported with multiple database searching.[2]

Search Strategy

The most important and often the most difficult aspect of an online search is the search strategy. Fenichel reports that the major problems are not found with the mechanics of the system command language but with the search strategy.[3] Search strategy includes selecting terms and synonyms which reflect the intended topic of the search, then determining the relationship between those terms in order for them to reflect the search request adequately.

Two important concepts of search strategy are *precision* and *recall*. Precision is a ratio measuring the number of relevant items retrieved to the total number of items retrieved. For example, if the searcher has retrieved 30 items from a file and 10 of these are relevant, precision is 10/30 or 1/3. Recall, on the other hand, is a measure of the number of relevant documents retrieved over the number of relevant documents in the file. If there are 20 relevant items in the file and the searcher has only found 10, recall is 10/20 or 1/2. If the patron is interested in a comprehensive search which will retrieve as many useful titles from the file as possible, a high-recall search will be performed. During the presearch interview, the searcher and patron need to decide the recall and precision levels to be sought.

Topics of recent, popular interest which find expression in common parlance with words such as *fast food* often result in searches of high precision, if word adjacency is often used to make the association between the two words [i.e., ? S *fast(w)food*]. Of the records retrieved in a search of *fast food*, the percentage that is relevant is likely to be high because of the unlikely possibility of a word association to be made in a context apart from the popular meaning of the expression. If *fast food* is given another, more formal name in the subject descriptors of a given database, a search conducted on that database only with the words *fast food* is likely to result in low recall. However, in cases where only a handful of relevant citations are needed by the patron, a high-precision, low-recall search would be useful.

Search Commands

In developing a search strategy, it is important to remember that the "computer searches for words, not concepts."[4] Words can be searched in a natural language fashion, in which the search analyst simply types the word or words in the manner in which they would normally appear in print (i.e., ? **S Backpacking**). The search for a word can also be narrowed to a specific field within the record, such as the title (? **S Backpacking/ti**), the abstract (? **S Backpacking/ab**), the descriptor field (? **S Backpacking/de**), or the identifier field (? **S Backpacking/id**). It is possible to search by author's name, as well.

In general, the subject content of a record is likely to be reflected in the title, abstract, descriptor, and identifier fields of a record. Combined, the subject-related fields compose the basic index of a database on DIALOG. Whenever a word is entered into a search statement (i.e., ? **S Backpacking**), without requesting a particular field, the search will be conducted for that word automatically, or by default, among all of the fields in the basic index on DIALOG. In DIALOG, searching the basic index corresponds to what is called *free-text* searching, which means essentially that nearly every word in the text of the subject-related fields may be searched.

When a search for a particular word is qualified to the descriptor field (i.e., ? **S Backpacking/de**), the search analyst is asking the computer to examine only that portion of the record in which the descriptors, or subject headings, for each record are found. These descriptors are chosen for each item by an indexer who must rely upon a thesaurus, or a list of controlled vocabulary, to select appropriate terms that describe each item.

The controlled vocabulary of a thesaurus is limited and is not able to account for the subject matter of all materials indexed in a given database. This shortcoming is overcome either through free-text searching or through the more narrowly defined search of the identifier field in a record. Words in the identifier field, like words in the descriptor field, are used to describe the subject content of an item. Unlike the words in a descriptor field, however, identifiers are not taken from a thesaurus. Identifiers are generally words taken directly from the title page or some other portion of an item which the indexer believes would be useful in describing the subject content of an item, and thus would be helpful in the retrieval of that item.

Each search option has its advantages and disadvantages, depending upon the type of material desired. A search utilizing controlled vocabulary identifies synonyms and draws related terms together, most often by the searcher's use of a printed thesaurus. In this way search terminology can be refined, and material located through descriptors has a high chance of being relevant and the search results more precise. However, inconsistent indexing exists, and it lowers the effectiveness of controlled vocabulary searching.

Natural language, free-text searching is useful for locating material on current topics, such as toxic shock syndrome. It is best used with highly specific topics, or for those subject areas that are not identified by a descriptor. However, there are disadvantages with free-text searching. Synonyms are not gathered, and so are not readily accessible, which many times decreases the precision level. Free-text searching may be misleading. For example, titles do not always reflect the actual meaning of the material. The title "Why Putt-Putt Isn't Sputter-Sputter," taken from the database ABI/INFORM, could be an article about golf, but in fact concerns how a U.S. firm is facing Japanese competition in the market for outboard motors. The use of free-text, or controlled, vocabulary depends upon the database, the topic, and the purpose of the search.

Other fields may be searched, such as publication year, language, document type, or subject code. These may be helpful in refining a search.

Boolean Logic

A patron is interested in finding any material on Soviet television. This request must be converted into a logical formula in order for it to be understood by the computer. There are two major concepts in this search: Soviet and television. These two concepts will be joined together and searched in the computer through a process called Boolean logic.

George Boole (1815–64), an English mathematician, developed a series of relationships that later became known as Boolean logic. It is a process which identifies whether an item is present or absent from a given set.

62 Technique

There are three forms of Boolean logic and each represents a different type of relationship. The OR operator requires that at least one of the concepts, or terms that represent a concept, be present in the retrieved citations. In an online search concerning the topic of Soviet television, the concept *Soviet* could be entered into the computer by the expression ? **S Soviet OR USSR OR Russia**. In this case, material will be retrieved which contains any of the three words. *Soviet* will retrieve material with "*Soviet* Union" and with "Union of *Soviet* Socialist Republics." If two or three of these terms are present in a single citation, the computer records the citation only once, anyway. This process can be represented by a Venn diagram (figure 1). The shaded area indicates that all material will be retrieved which contains the words *Soviet* or *USSR* or *Russia*.

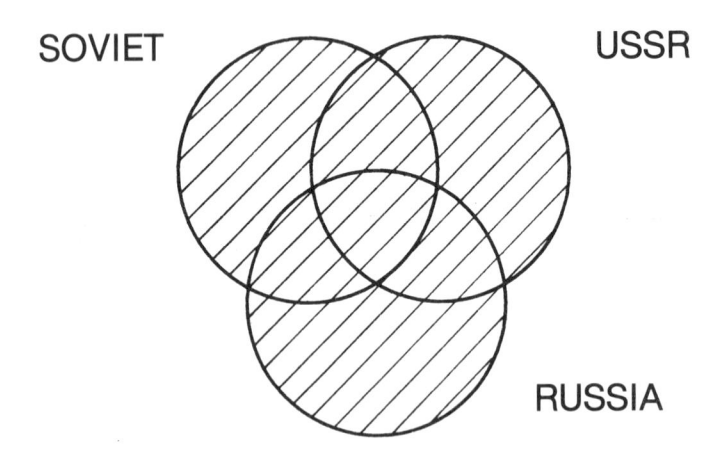

FIGURE 1. Venn diagram illustrating OR operator

The OR operator can also be used to combine synonyms of the television concept. For example, the statement "television OR TV OR mass media" will retrieve all of those materials that contain the words *television* or *TV* or *mass media*.

The second Boolean operator is the AND operator which requires both concepts to be present. In the Soviet television example, the patron is interested in materials containing the concepts *Soviet* and *television*. If an item only contains the concept *television*, it would not, for all practical purposes, be as useful as an item that contains both concepts. The use of the AND operator means that both terms *Soviet* and *television* must appear in the item for it to be retrieved. An example of the results of using the AND operator is illustrated in figure 2.

SOVIET   TELEVISION

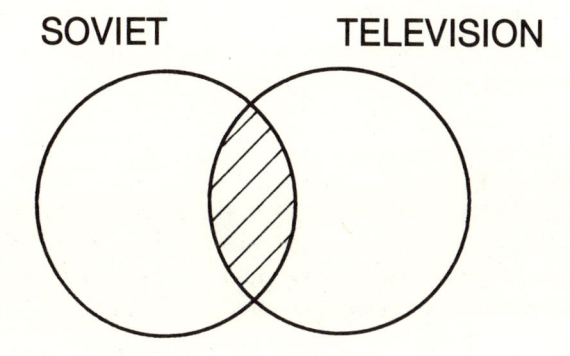

FIGURE 2. Venn diagram illustrating AND operator

The third Boolean process, the NOT operator, is used to subtract a concept from a set. The patron in our Soviet television search may be interested in television, but not at all interested in radio. In the example, the searcher could use the NOT operator to exclude all of the materials that mention *radio* by the statement "? **S Television NOT Radio**." Figure 3 illustrates this example. However, notice that material containing both the words *television* and *radio* are not included in the final set. Therefore, caution is advised when using the NOT operator. Many relevant citations may have been lost that deal with both Soviet radio and television.

TELEVISION   RADIO

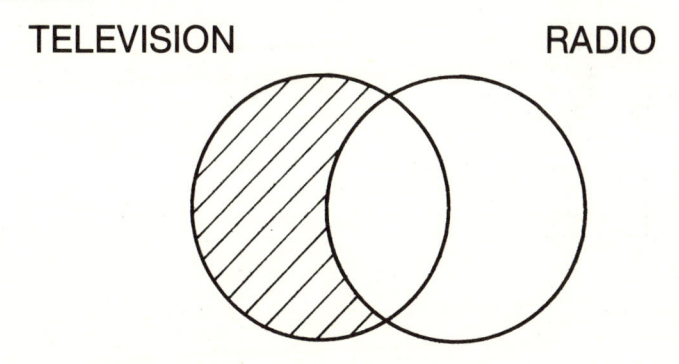

FIGURE 3. Venn diagram illustrating NOT operator

Boolean logic is a process by which requests are formulated into a series of relationships that will be understood by the computer. This approach also aids the searcher in developing a search strategy which reflects the patron's request. However, Boolean logic does not guarantee that all items retrieved will be relevant. The computer only retrieves material dealing with Soviet and television, but that does not necessarily mean the two concepts will be related as the patron desires in the materials retrieved.

Strategy Approaches

Given a particular search request, there are many different strategies to use. There is no perfect search formula, but three main search strategy approaches exist which may guide the searcher. The three models are building blocks, successive fractions, and citation pearl growing.[5]

Building blocks. The building-blocks method, illustrated by figure 4, is the predominant strategy for many searchers. Through this approach, each concept of the search is developed separately, then all concepts are combined to produce the desired results. The Soviet television search is an example of the building blocks method. First, the concept *Soviet* is defined using the OR operator, then the concept *television* is defined, also using the OR operator. These two groups are combined with the AND operator and the results are printed and analyzed for relevancy.

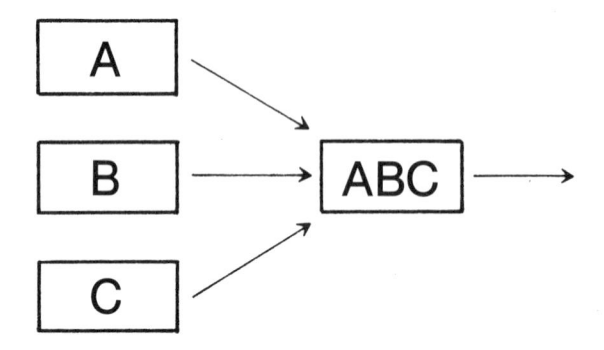

FIGURE 4. Building-blocks method

This actual search will be demonstrated using the PAIS database through the DIALOG retrieval system. However, before this is illustrated, search command language will be explained. The "prompt" in the DIALOG system is a question mark (?). When it appears, the computer is ready to receive the information from the searcher.

The "SELECT" or "S" command is used when the searcher wants the computer to find or select items that contain a particular word or phrase. For the Soviet television search, the first concept to retrieve, using the building-blocks approach, is the concept *Soviet*. Therefore, the following is input into the system (boldface indicates that an item was put into the system by the searcher):

? **S Soviet OR USSR OR Russia?**

The question mark after the word *Russia* will retrieve materials containing words beginning with the six letters "r-u-s-s-i-a," including *Russian* or *Russia's*. This sort of retrieval is an example of truncation.

The computer responds by indicating the number of materials containing the words *Soviet* or *USSR* or *Russia*:

        1647 Soviet
        283 USSR
        4341 Russia?
1   4995 Soviet or USSR or Russia?

(The numeral one indicates this was the first set of citations retrieved.)

There are 4995 records containing the word or words *Soviet* or *USSR* or *Russia*. This total is not the sum of the three numbers because duplication of materials has been eliminated by the search.

The same approach is used to retrieve materials for the second concept, *television*. (The "w" between *mass* and *media* calls for the retrieval only of the materials in which the words *mass* and *media* are next to each other, i.e., not articles in which these words are found separated by other words. Without the "w," the computer limits the search to multiple words in controlled vocabulary fields.)

? **S televis? OR TV OR mass(w)media**
        811 televis?
        259 TV
        350 mass(w)media
2   1168 televis? or TV or mass(w)media

There are 1168 records containing the concept *television*.

The next step, using the building-blocks method is the combination of the Soviet and television concepts. The Boolean operator AND is used to retrieve items that contain both of the concepts *Soviet* and *television*. The "COMBINE" or "C" command is used when the searcher combines the results of previous search statements through the use of the number by which they are identified. The first search statement deals with the concept *Soviet* (set no. 1) and the second with *television* (set no. 2). These two are combined by:

? **C 1 and 2**
        3   13 1 and 2

The result is a third set statement with 13 records retrieved from the combination of sets 1 and 2. The combination of sets 1 and 2 has yielded only 13 items which contain the words that were sought in both sets 1 and 2. These 13 items are arranged in set 3 in numerical order by the accession number of each item, from the latest number to the earliest. (The accession number is applied to each item as it is put into the database.) With the "type" command, the searcher is able to view any or all of the 13 items in set 3, and in any specified order. To see the first 2 items, the searcher types in a command as follows:

    type   set number/format/items

or:

    ? type 3/3/1-2

3/3/1
The Media in a Soviet Industrial City.
Soviet Sociol 20:28-52 Summer '81.

3/3/2
The Mass Media and Public Communication in the USSR.
Remington, Thomas
J Politics, 43:803-17 Aug '81.

The portion of each record that is the bibliographic citation is retrieved through the use of DIALOG type-format 3. Other formats retrieve the title alone, the full citation plus an abstract, etc. From this printout, the user is able to locate the actual article on Soviet television in the journal *Soviet Sociology* in the library.

The building-blocks approach develops each concept separately, and these results are combined to produce the desired set. This approach is clear and logical; therefore, backtracking to revise the strategy is a relatively easy process.

> *Successive fractions.* The second method, successive fractions, is best used when the search topic is broad or vague. A broad topic is entered into the computer, and the searcher is able to chip away at this set until a desired number of results are retrieved. This chipping away process may include limiting by document type, year, or language, or may also include limiting the search to titles in citations only, and not searching for them in abstracts, etc. The successive-fractions approach is illustrated in figure 5.

A patron is interested in finding material on mainstreaming in secondary education. The searcher decides to use the ERIC database and

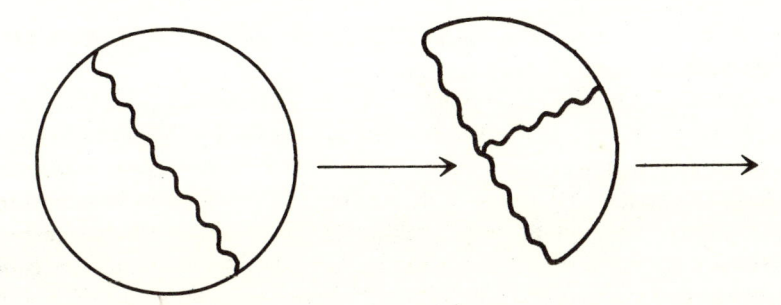

FIGURE 5. Successive-fractions method

enters the following phrase to retrieve material pertaining to both topics:

? S mainstreaming/df AND secondary education
        2630 mainstreaming/df
        32172 secondary education
1   273 mainstreaming/df and secondary education

In this example, both concepts have been input as descriptors. The "df" attached to *mainstreaming* identifies it as a single-word descriptor. *Secondary education* without a "w" separating the two words designates it as a multiple-word descriptor.

Even though all 273 references may be relevant to the search request, it may be time-consuming to sort through each citation. The set can be reduced by document type to include only material written as journal articles.

? Limit 1/EJ
   2   99 1/EJ

Only 99 of the original 273 items are journal articles (designated by "EJ").

The set can also be restricted by year. If the patron were only interested in material written since 1980, the following command, limiting by accession number (which is also denoted by "EJ"), would be entered:

? Limit 2/207485-999999/EJ
   3   54   2/207485-999999/EJ

Strictly speaking, EJ 207485 will appear in the 1980 index, but it may have been written before 1980. Accession numbers are assigned to materials as they are put into the database, but these numbers do not necessarily indicate when the material was published. Restricting materials by accession numbers is one way to limit by year. (The range of accession numbers for materials put into ERIC by year can be found

68 Technique

through the use of the DIALOG "EXPLAIN" command, "? ?LIMIT1," which will give all of the available uses for the limit command for DIALOG file 1, which is the database ERIC.)

The set which originally began with 273 references has now been reduced to 54 journal article citations entered into ERIC after 1979. The set has been reduced to a workable size by partitioning the broad search topic. Search revision or backtracking is easier using the successive-fractions strategy rather than the building-blocks approach because single limiting concepts can be subtracted with one command. This type of search is completed when the desired number of items are retrieved.

> *Citation pearl growing.* The third search strategy, citation pearl growing, involves retrieving a record already known to be relevant in order to identify subject descriptors that specifically identify that relevant record. These descriptors are used in a search strategy to retrieve additional relevant materials. The citation pearl-growing process is illustrated in figure 6.

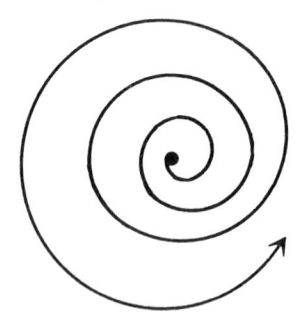

Reprinted from Robert E. Buntrock, "The Effect of the Searching Environment on Search Performance," *Online* 3 (October 1979):12.

FIGURE 6. Citation pearl-growing method

To illustrate, a patron is interested in locating material dealing with employee fitness programs and employee performance. A relevant citation is brought to the presearch interview:

Edwards, Sandra E. "The Effect of Employee Physical Fitness on Job Performance." *Personnel Administrator* 25 (November 1980): 41–44.

The searcher is able to locate this article in the database ABI/INFORM by conducting a search on the author's name.

? S AU=Edwards, S?

One citation is discovered that has "Edwards, S" in the author field:

1   1 AU-Edwards, S?

The searcher enters the "TYPE" command (abbreviated as "T") for the first set. Type-format 2 on DIALOG will display the entire record except the abstract.

? T1/2/1
1/2/1
80021703   I.D. No: 80021703
   The Effect of Employee Physical Fitness on Job Performance
   Edwards, Sandra E.; Gettman, Larry R.
   Personnel Administrator v25n11 41-44,61 Nov 1980 Coden: PEADAY ISSN 0031-5729 Jrnl Code: PAD
   Availability: ABI/INFORM
   Doc Type: Journal Paper
   Descriptors: Employees; Performance; Physical fitness; Exercise; Impacts; Studies; Job satisfaction; Effectiveness.

The searcher and the patron decide that the main concepts that are sought appear in the descriptors *job satisfaction* and *physical fitness*. The searcher is now ready to group these two concepts together in order to retrieve relevant citations.

   ? S job satisfaction
        2   1079 job satisfaction
   ? S physical fitness
        3   150 physical fitness
   ? C 2 and 3
        4   2 2 and 3

Two references satisfy the requirements of the above sets. The searcher is able to scan the titles and descriptors of the two items by using DIALOG type-format 8 in order to make sure the items are relevant.

   ? T4/8/1-2
4/8/1
82021247   I.D. No: 82021247
   Employers' Perceptions of Benefits Accrued from Physical Fitness Programs
   Descriptors: Employee benefits; Physical fitness; Programs; Exercise; Personnel management; Perceptions; Models; Job satisfaction.

4/8/2
80021703   I.D. No: 80021703
   The Effect of Employee Physical Fitness on Job Performance
   Descriptors: Employees; Performance; Physical fitness; Exercise; Impacts; Studies; Job satisfaction; Effectiveness.

The two items above are relevant to the search request. The citation pearl-growing approach can be executed again, using new descriptors or free-text terms drawn from the two new citations in order to retrieve additional, potentially useful, citations.

The citation pearl-growing strategy uses the interactive capabilities of the computer in an efficient manner. This approach may take a longer time online than the other two strategies, but the advantages makes this strategy unavoidable in a number of cases.

Online/Offline Printing

There are two ways to view the results of a computerized database search. Online printing results when the searcher wishes to view the types of materials retrieved from the search while still being connected to the computer. Online printing enables the searcher and user to view the results instantaneously and thereby revise the search, if necessary. Online printing can be expensive if many citations are displayed, as the cost of the search is determined by the charge per connect hour.

Offline printing involves typing the "PRINT" command (rather than the "TYPE" command) in order for search results to be compiled at the computer center and mailed to the searching institution. Offline print results arrive approximately one week after the search has been executed. Offline prints are less expensive than online prints if there is a large quantity of information to retrieve, but the disadvantage of offline printing is the delay. Many users may not be able to wait a week for search results.

Post-search

Results of a search are useless if the patron is unable to interpret the printout. After the search is performed, the search analyst should explain any notations that are unclear to the user. If the material is not available in the library, interlibrary loan and vendor document ordering services can be utilized.

Conclusion

This paper outlines the mechanics of online searching for the novice searcher. Database selection, Boolean logic, and search strategy have been discussed, as well as printing options. Post-search explanation is necessary in order for the user to understand the results.

Notes

1. Carol H. Fenichel and Thomas H. Hogan, *Online Searching: A Primer* (Marlton, N.J.: Learned Information, 1981), pp. 12–14.
2. Donald T. Hawkins, "Multiple Database Searching: Techniques and Pitfalls," *Online* 2 (April 1978): 9.
3. Carol Hansen Fenichel, "Online Searching: Measures That Discriminate among Users with Different Types of Experience," *Journal of the American Society for Information Science* 32 (January 1981): 23.
4. Arthur L. Adams, "Planning Search Strategies for Maximum Retrieval from Bibliographic Database," *Online Review* 3 (December 1979): 373.
5. Karen Markey and Pauline Atherton, *ONTAP: Online Training and Practice Manual for ERIC Database Searchers* (Syracuse, N.Y.: ERIC Clearinghouse on Information Resources, Syracuse University, June 1978), pp. 17–24. Based on the earlier unpublished work of Charles Bourne.

Part 2
MANAGEMENT

PETER G. WATSON

Library Organizational Patterns in Online Retrieval Services

This paper will consider the organizational framework for online retrieval services in libraries. The discussion will be focused on the pros and cons of two major concerns: first, whether to establish the service as a centralized or a decentralized operation, and second, whether to integrate it with existing reference service or create a separate unit for computerized searching. We will conclude with some observations on the key middle role of the service coordinator between the searchers and the library administrators.

It is clearly essential for the organization to agree on how any new service will be offered, and to know what relation the new service will have to the library's existing service posture. To state these two points is to establish the overall context within which such fundamental management concerns as the nature, cost, staffing, and even the style of the new service can be addressed. Decisions in all of these areas will influence the library-user community's acceptance of this information service.

However, opening day certainly does not, and should not, signal the end of all thinking about the organizational framework of the service. It is absolutely vital, particularly in these times when organizational structures are not considered as fixed and immutable as in an earlier age, that the original administrative choices be reviewed, at least every few years. Experience provides a basis for a different perspective on the original decisions: one can now ask not merely Which course shall we adopt, X or Y? but How is X working out, and how might Y be an improvement, in light of what we now know about X?

Centralized vs. Dispersed Authority

The problem of centralized versus dispersed authority for the online service in one sense can be seen as a variant of that much more general question in library administration, the problem of one library or several branches. This has had particular relevance for academic libraries in recent years, most notably in the case of the undergraduate library, but it has also faced library managers and administrators in special and public libraries. In some situations, there is really no choice: a large, geographically dispersed national corporation may already have several fairly autonomous libraries operating in various parts of the country, or indeed the world. Public library systems in metropolitan areas, or in counties, tend to have many branches, but a strong, centralized administrative structure. The branches are mainly additional service outlets for basic circulation of materials, with most of the collection development, ordering, processing, and the more difficult types of reference service usually handled centrally. As an example of decentralization, the branch-library arrangement is a very different thing from, say, a large, almost autonomous medical or law library on an Association of Research Libraries university campus. So it is very likely that the pattern of management and administration adopted for the placement of computerized search services will mirror, and will have to fit with, what already exists. There are several main options:

1. A single, centralized service.
2. A dispersed service having strong local management and a deliberately weak central coordination. I employ "weak" here in the political-science sense, the sense in which it is said that some cities have a weak mayor's office, but a strong city council, or vice versa.
3. A dispersed service having a strong central administration. Searching is done at the service points, but publicity, billing, accounting, policy setting, vendor relations, training and other management functions are controlled centrally.
4. A hybrid pattern. This works very well for the library system of Bell Laboratories, for example, whose director Donald Hawkins has given an excellent overview of the major factors in managing an online service:

> At Bell Laboratories, a hybrid mode of operation is very successful. Local online service, fully integrated with standard reference service, is provided in each of the units of the Bell Laboratories Library Network. In addition, a centralized service staffed by three Ph.D. Information Scientists, performs sophisticated, time-consuming or other specialized searches, and acts not only as a backup to the local services, but also as a consultant to the Library Network on all aspects of online

searching. . . . Of course, only a few organizations have either the financial resources or the searching volume to justify such a depth of staffing in one area.[1]

This may serve as the epitome of computerized searching under favorable conditions: a large clientele of goal-oriented researchers, a high caliber search staff, a large corporate budget commitment, and a routine acceptance of computerized searching as a normal part of library service.

Whatever arrangement is eventually adopted, it is wise for an institution to establish one final and well-understood source of authority for budgetary and policy matters. A high degree of local autonomy is fine, but if it extends to one unit offering online services according to totally different policies and a cost system different from the other units, then autonomy begins to be counterproductive. Patrons rapidly notice such differences and will soon start to "shop around" for the best bargains. They will probably also start to suspect that there are irreconcilable organizational differences within the library system, which, whether true or not, is detrimental to the library's image. And a further effect is felt by the librarians themselves. Those not fortunate enough to work in the unit which is going its own way will quite likely come to feel like second-class citizens, with corresponding effects on their behavior. So while central coordination of a group of searching units can be weak at the operational level, overall administrative direction certainly should not be.

Of course, in addition to the existing type of library structure, a further factor in the decision to centralize or disperse is money. Maybe you would *like* self-contained search stations in every branch, but maybe you cannot afford all that remodeling and equipment cost. A decentralized operation will, in all probability, cost more than a centralized service of comparable scope. For a single search operation in a multi-unit library structure, you would also be choosing a separate, rather than an integrated, service with at least some of your patrons, since searching is being done by staff other than the reference staff in the branches.

Separate or Integrated Service

ADVANTAGES OF INTEGRATION

1. The same reference staff can provide a full spectrum of reference services. Put another way, they will have a greater range of tools available from which to offer good library service. All types of resources can be deployed in an integrated fashion. An inter-

esting corollary to this, the stringent fiscal constraints under which most such services operate being kept in mind, is that information requests which should *not* become online searches can probably be diverted more efficiently into other sources.
2. Staff are available to answer questions about the service, and perhaps initiate a search, for more total hours per week, since they will merge this function into general reference desk activities.
3. This approach avoids competition between a reference operation and a search service for budgetary support, space, and, one may add, status.
4. Related to the above, an integrated approach will foster an enhanced sense of professionalism among the reference librarians who become searchers, and perhaps, by the halo effect, among others in the department also.
5. This in turn is likely to affect the user's perception of what the librarian is and does. Since the patrons are in some way always related to the funding of the library (whether as taxpayers, in a public library, or as clients on whose vote of confidence the library ultimately relies, as in an academic or corporate library) the value to a library of a new group of patrons who are thus induced to feel proud of "my library" cannot be overstated.
6. New practices of the online reference service can trickle down into regular reference service. Two examples are new techniques of questioning the patron, and new ways of explaining a particular reference tool. Northwestern University has recently formalized one such practice by initiating an appointment system for other types of reference transactions not merely the online search.
7. The same clerical staff can probably deal with the needs of the online search service, among the most important of which is the need to have a support staff which can handle at least the first level of patron inquiries about how to obtain a search, how to pick up the output, where to pay any charges, etc.
8. Integrating the service into reference is probably less hard on any who prove unsuccessful as searchers. If they are reference librarians first and foremost, then the computerized version of reference work is not their whole life, and they can have a useful and fulfilling role in the reference operation without any need for traumatic solutions such as being transferred out of a section which is wholly dedicated to searching.
9. In many academic libraries (and, one would hope, in other types of libraries too) reference librarians are heavily involved in collection development activities. Thus, changes in collection development policies as a result of the online searching process

can be both suggested by the searcher and also acted on by the same person. Some examples are the need to start buying serials in a new area or a decision about binding back issues or obtaining microform. In general, a person in this combined role knows much more about what patrons are using; in some libraries the searcher routinely keeps a copy of the search output by having two-part paper in the printer.
10. An integrated service, as implied earlier, is probably more feasible in those institutions (and there are many) which, to begin with, can designate one, or at most two, searchers. We have already mentioned budgetary considerations, and clearly the additional overhead involved in establishing a whole new section for just one or two searchers would usually be prohibitive.

ADVANTAGES OF A SEPARATE UNIT

If your institution has a slant toward programmatically based organization, by virtue of strong grants and contract activity, for example, then the separate-unit approach might be an appropriate pattern to adopt. After all, data services are in effect a new program, whether submerged in a reference operation or not. Form follows function, and in other respects the library has, over time, adapted its organizational structure to give identity and recognition to new services or functions.

1. Better visibility. A separate unit will have identified personnel with clear job titles, a cohesive function, and even a separate location with a "shingle on the door." This visibility provides a focal point to which patrons can relate.
2. Greater operational autonomy. The unit will usually have an expert in charge, namely, the library's most experienced searcher (who almost always will have spent at least some time in general reference work). Thus, there is more scope for informed decisions about how the service is run, for example:
    How it is publicized to patrons
    What is and is not offered (which services, databases, etc.)
    Selection of searchers
    Training of searchers
    Budgetary control (one single, programmatic budget)
    Secretarial support.
3. There will be a better chance of gathering together a highly motivated staff; it is *their* operation, *their* performance, which matter. A separate unit can foster the feeling that the success of the enterprise depends only on them and is not so much at the

mercy of complex interrelationships, or of issues whose resolution is based more on organizational politics than on merit.
4. There will probably be a better chance for growth, once such a separate unit is established and allowed to adopt its own natural form. For example, justifying a new search position may be an easier thing to do than justifying a new reference librarian position. Planning for growth is several degrees easier.
5. A separate unit will have the ability, other things being equal, to respond faster to needed changes in the services offered and in the way they are offered. There will be fewer people to convince, fewer approvals to gain, fewer "environmental impacts" to thrash out before being able to act.
6. There will very likely be more technical efficiency in the search process, because of the higher level of skill that comes with having full-time, or nearly full-time, people. Searchers will be more likely to achieve general, not just subject-specific, mastery of what they are doing; they will become generally familiar with the nature of systems, not just of given databases. Often, full-time search positions will attract people with higher educational qualifications —in fact, such positions can be deliberately structured to do so (cf. Bell Laboratories).

The disadvantages of each of these broad organizational patterns (for that is all they are intended to be) can be seen from what has been said of their respective advantages. The arguments for one become the arguments against the other. For example, if we say that integrating the search service with other reference operations enables a librarian to utilize all available information tools, we imply that this will not be true to the same extent in a separate operation, and that this lack is a disadvantage. If we say that a separate unit will usually have the ability to respond more quickly to needed changes, we mean by comparison that an integrated service will have a greater number of competing claims on a reference librarian's time which may, indeed, become a drawback. From this, the counter-arguments can be built up, and managers or administrators can attach to both the advantages and disadvantages of each some type of weighting that best fits their own organizational circumstances.

Data Services Coordinator

Whatever the model eventually adopted, it is likely, judging from experience, that there will be a need for someone with a title such as data services coordinator, who is, depending upon the organizational climate in which the position exists, a team leader of the search librarians, a planner, an advisor to the administration, a focal point for contact with the outside world, a troubleshooter, and an evangelist. Let me give you an example of the central position occupied by the data services coordinator: I once identified about a dozen groups of people, inside the library and out, with whom I dealt, and to whom I was the primary representative of this new service. As a coordinator on a campus that belongs to a multi-campus university system, I listed: library patrons; library colleagues; library administrators; campus administrators; campus staff officials (e.g., the purchasing office, the personnel office); campus computing center staff; university system-wide staff who had responsibility for libraries and their programs; other campus libraries in the system; information service vendors (e.g., DIALOG, BRS, SDC); equipment vendors; database manufacturers; telecommunication specialists (a campus office and/or the phone company); and other outside groups or individuals (e.g., campus visitors, local journalists, etc.).

The library world will, I believe, eventually recognize what it owes to the relatively small number of people who have filled this role. Most frequently they have been working reference librarians, intelligent, enthusiastic, for the most part fairly young, and not lacking in other tasks that must be done during the normal work week, but dedicated to the idea that these new methods of information service must, for the public good, find a place among the functions of the library; a proposition now largely accepted, I am pleased to say.

The data services coordinator is, then, a service specialist who perforce became something of a technical specialist (at least in relative terms) in the field of computer-based searching. But the coordinator is also, and most importantly, a distinct figure in a social system—the agent of change. As such, the coordinator is, or soon learns to be, something of a peer-group leader, a missionary, a persuader, an advisor, and a "go-fer" when, as is often the case, the service would suffer if new paper was not put in the printer or file update pages not entered into system manuals. The coordinator is usually responsible for seeing to it that a service is provided and needs at least a part of the time of everyone who is to be a searcher, but generally cannot command the time and the participation of anyone. In many instances, the data services coordinator is asked to take on the stressful task of starting and operating a computerized search service in addition to all the other duties of a full-time job, and can therefore often be seen in the library after hours, or at odd times during the weekend. Not infrequently, such quiet times are used to do a personal searchload, for the coordinator is customarily also a designated searcher in some given area, or will have searches to do, in the role of chief backup, in case of illness in the group, etc. It is a busy, overworked existence, but it brings the tremendous satisfaction of having accomplished goals to which all public service librarians are profoundly committed, by a new, hitherto untried, technically challenging, but clearly necessary method, the computer.

Note

1. Donald T. Hawkins, "Management of an On-line Information Retrieval Service" in *The Library and Information Manager's Guide to On-line Services*, Ryan E. Hoover, ed. (White Plains, N.Y.: Knowledge Industry Publications, 1980), 99.

RANDOLPH E. HOCK

Who Should Search?: The Attributes of a Good Searcher

Who should search? I will confess that this has long been an adrenalin-producing topic for me. There are several reasons why I so respond to the topic. One is, that as a librarian (or at least a former librarian), I feel very strongly that the future of reference librarianship is going to be determined to a large degree by the use of online systems, that the quality of reference service will be largely dependent upon the quality of searching, and that the quality of searching will be largely dependent upon *who* is searching.

The second reason for my interest in the topic is that my largest present professional responsibility is the training of searchers. Having personally trained over 2,000 searchers, I find that the qualifications, personalities, and backgrounds that students bring to the world of searching are extremely diverse. Some of the qualifications are more predictive of success than others, and, indeed, there are some people who simply will never make good searchers. For those called on to select searchers, or who are asking themselves if they should begin searching, an examination of these factors should be instructive.

The issue of who should search is becoming especially important with the advent of the end-user searcher. More and more end users are doing their own searching; among the most frequently stated reasons for this is that they either cannot get searching done by their library or they are dissatisfied with the results they are getting from librarians. I strongly feel that the best searching can be done by librarians. But if librarians won't or don't search well, someone else will search.

The opinions expressed in this paper are the opinions of the author and do not necessarily reflect an official position of DIALOG Information Services.

Attributes of a Good Searcher

A very strong case can be made that the quality of the searcher is the most significant element in the quality of a search. Therefore, it is not just important, but imperative, to examine what qualifications are necessary to make a good searcher, and in selecting searchers, to have some feel for what factors are necessary for successful searching. The factors I would identify fall into two categories: personality and background of the searcher. In the personality list are the following: (1) a logical, analytical mind, (2) communications skills, (3) enthusiasm, (4) economic attitude, (5) courage, (6) ability to make quick decisions, (7) being a good student, and (8) a rudimentary typing ability. Under background are reference training or experience and subject expertise.

Now to go into detail on the list of personality factors: First, a logical and analytical mind. This is required above all other factors, and we must recognize that not all people are equally endowed with this. By analytical, I mean analytical in its most precise sense—the separation of a whole into its component parts, the ability to look at a problem and quickly say, "We are dealing with concept A, concept B, and concept C. All other aspects of the problem can, for the moment, be put aside while we tackle these most important parts." The ability to put the lesser ideas aside is surprisingly rare. It is entirely possible that many beginners' searches are failures because they overspecify the topic and become bogged down by less important ideas.

Another aspect of the logical, analytical mind is the ability to deal with Boolean logic. Boolean logic is, by most standards, not difficult, but not everyone can easily handle it. A prospective searcher should check out a short text on set theory and try it out before making the commitment to searching.

The second personality factor on the list is communications skills which include interview techniques, listening ability, the appropriate level of assertiveness, and, in some contexts, public-speaking skills. These are necessary, of course, as part of the reference interview stage of the search process. The searcher must have the facility to restate the same ideas in two or three ways to be sure of communicating with the requester. The searcher must have the ability to listen, to hear not just the words, but to hear, as we said in the sixties, "where the person is coming from." The searcher must ascertain the technicality the person requires, the exhaustiveness the person seems to need, what the requester actually knows about the topic, and so on.

Assertiveness is a critical element in the area of communications skills. The searcher must be able to imply tactfully that, although the requester certainly knows the topic better, the searcher is the one who knows how to search. A well-organized searcher who can ask the right questions and knows the appropriate techniques and databases can generally convey this without being too blunt. Does the prospective searcher find it easy to tell psychology graduate students that, in spite of their having asked for the location of *Reader's Guide*, they very possibly need *Psychological Abstracts*, or does the prospective searcher obediently lead the student to the *Reader's Guide* shelf? This assertiveness is especially important when the requester will be sitting at the terminal during the search, because an inadequately aggressive searcher can easily be led quite a few dollars off the topic.

Public speaking skills are not always necessary but, on the other hand, are sometimes required as part of the searcher's job. If promotion of online services is part of the job, skill in speaking to groups can sometimes make the difference between success and failure of the service. This ties into the third of the personality factors, enthusiasm.

If the requester gets the impression that searches are a painful experience, you may lose not just an online patron, but a library patron. Most of you, I am sure, find that increasing numbers of your patrons are already aware of online services and know that good results are indeed attainable. If the user encounters a librarian who is enthusiastic about online searching, this can reflect on the services of the library as a whole. The enthusiasm qualification is another qualification which is fairly easily determined by observation.

The fourth personality qualification mentioned was an appropriate economic attitude. This is somewhat more difficult to identify in the prospective searcher, and also somewhat more difficult to define. Perhaps a few examples will help. First, I find that it is very easy to be penny-wise and pound-foolish online. A searcher must recognize that spending two or three more minutes online comparing different approaches to a question can sometimes change a virtually worthless search into a good search. At the same time, a searcher must be able to recognize that there are some situations where it is better to settle for a larger portion of irrelevant citations than to double the online time refining the results. Since pricing structures for some databases are becoming more complicated, searchers should, while online, be able to mentally juggle figures quickly so that they have, without too much conscious effort, a continuing feel for how much is being spent. On the other hand, there are some searchers who seem to spend more of their online time thinking about costs than they do about the search itself. Finally, the searcher should be astute enough, when comparing prices among systems, to be able to figure out whether a 5 percent lower connect rate is a real savings if it means access to a database that is 30 percent smaller.

The next qualification, courage, I was unsure about listing. I decided to include it though, because although the term was admittedly chosen with a degree of hyperbole, I did want to point out that timidity, with respect to either the terminal or patrons, can ruin the chances for good searches. I would guess that all of us, the first time we used a terminal, had some fear that we could easily destroy either the terminal, the system, or ourselves. I think this is very natural. If it persists, though, the searcher is always going to be too concerned with making mistakes and consequently will make them, since there are a lot of other things the searcher at the terminal really should be thinking about. Patrons sense these fears and can easily lose confidence in a searcher.

Sixth in the list of personality factors is the ability to make quick decisions. Although the appropriate level of preparation can obviate many otherwise quick decisions, the reasons for this as a qualification for a good searcher should be fairly obvious. I find that good searchers don't always have to rely on snap decisions, but develop a knack for being able to pull up the next item of information on the terminal while they are digesting and deciding on the present item. The searcher does need to be constantly reformulating the list of options and contingencies. The ability to do this quickly can make a difference not just in terms of the efficiency of the search but in its effectiveness.

As the seventh personality factor I listed being a good student. Searching is a matter of continual learning. For the beginning searcher it means sitting in classes, taking notes, conceptualizing, remembering, but most important—enjoying learning. I think whether or not the prospective searcher participates in other noncompulsory classes is a good indication of whether the person will meet this qualification.

The last factor I list is typing ability. It is placed last on purpose. I think it has some importance, but is of much less importance than the other factors. One does not have to be a good or even mediocre typist in order to search. If, on the other hand, one has to spend several seconds looking for the "s" on the keyboard, then obviously a lot of valuable time will be wasted. A convenient yardstick is this: If sometime when you were in college you managed to type a few papers of your own, your typing ability is probably adequate for searching.

The second of the two major categories of qualifications is background, including more specifically reference training or experience and subject expertise. A reference background is necessary because the purpose of online systems is not for searching as an end in itself. Online systems are rather a way of making use of a large collection of research tools. Online systems can radically increase the usefulness and accessibility of those tools, but in the end it is the research tools, the abstracting and indexing services, the directories, the statistical collections, which are being accessed. If the searcher does not know what tools exist, what kinds of information are to be found in those tools, and how the tools differ from one another, then those tools cannot be used to their full advantage. It is for this reason, above all, that the skills possessed by a good librarian or information specialist are essential if an organization expects to achieve maximum use of online systems.

The second background qualification, subject expertise, is probably the most controversial of all the qualifications on this list and is, quite frankly, an issue for which there is probably no final resolution. The degree of subject expertise required is mitigated by the personality and talents of the individual searcher and varies according to the subject involved. The optimal situation is to have a searcher who is both a subject specialist and a reference librarian. There are, indeed, some large corporations in which searchers are both. More frequently though, because of the breadth of searching being done, the searcher is not an expert in all fields searched. Some trade-offs are usually required. The most general statement possible is that, given a reference background, the more the searcher knows about the subject, the better.

There are a few areas in which there is almost a consensus that at least some subject background is necessary. On this list, chemistry is probably the most demanding subject background, followed by biomedicine and the other sciences. Depending on the nature of the searching environment, the list might also include patents and financial searching. In the case of chemistry, at the very least some acquaintance with chemical nomenclature is essential. If the searcher does not know how to spell *methyl* or does not know that alkenes and olefins are the same, then good-quality chemical searching will be almost impossible. The chemist requesting the search is going to figure that the search might as well be done by a chemist rather than a librarian.

For any subject area, the less specialized the language of that field, the greater the likelihood of success for the nonspecialist searcher. The problem is basically one of communication. The better the searcher can communicate with the requester, the better the quality of the search.

Selecting Searchers

For those who are called upon to choose searchers, though, there is one further issue: how many people should be searching. It should be remembered that high-quality searching is a skill which requires experience. The more experience searchers get, the better their searching should be. If an organization is doing thirty searches per month but divides those searches among six searchers, none of the searchers is likely to gain enough experience to search with ease. In most cases, a searcher who is only doing one or two searches per week is probably spending a significant amount of time at the terminal relearning. This is not the fault of the searcher, and may, indeed, be an unavoidable situation. It should be remembered, however, that online experience is as important as online training. Whatever can reasonably be done to increase the level of experience of the searchers should be done, and in many situations this could be accomplished by having only one or two searchers trained, instead of several.

Finally, a word should be said about the gravity and distribution of the responsibility for selecting searchers. As terminals and personal computers proliferate, the tendency for the searching function to slip away from the librarian/information specialist is increasing. Though this may be in some cases advisable, it can also be in the end detrimental to the end user and the end-user's organization, perhaps even to the functioning of the information infrastructure of our society as a whole. If this is to be avoided, we must get the best people to the terminal. Administrators must recognize that choosing searchers has to be done with extreme care. They must recognize also that in order to get the best searchers, those people must be adequately compensated, even if it means changing traditional salary structures.

Library schools also share the responsibility. Even more than in the past, they need to be selective of candidates, with an eye both to the probability that their graduates will be doing online searching and to make provisions for the searching positions in which subject expertise is necessary. Library schools must make sure that an appropriate proportion of graduates have the necessary subject backgrounds.

Finally, the responsibility for choice of searchers rests also with the prospective searchers. Searching is an activity which can be interesting, fulfilling, and enjoyable. That the searcher enjoys searching might, in the end, be the most telling qualification of all.

M. R. DUSTIN

Library Applications of Database Searching

This paper will provide an overview of three areas in which database searching can be implemented in a library. While advantages and disadvantages of online searching in libraries were thoroughly discussed by Jane Thesing earlier in this volume, this paper gives actual examples of how some of the advantages work. The three areas chosen, suggested by "An Introduction to Online Searching," the appendix to this book, are:

1. The development of a bibliography of sources, often annotated, which can be quickly and comprehensively prepared on a wide variety of subjects
2. The use of nonbibliographic or source databases which provide primarily, at this time, business and statistical data
3. A variety of ready-reference uses which directly provide answers to some types of questions, especially those requiring either very current sources of information or bibliographic verification.

The examples are actual questions that have been referred to the Minnesota Interlibrary Telecommunications Exchange (MINITEX) reference service by participating public, academic, and state agency libraries. MINITEX was established as a program of the Minnesota Higher Education Coordinating Board to facilitate resource sharing among participating libraries. The reference service is one of the five activities designed to carry out this objective. Questions referred to us are ones which cannot be answered locally with existing resources. To locate requested information we make use of online databases, a WATS telephone line, the resources at the University of Minnesota Libraries and Minneapolis Public Library, as well as the expertise and collections of other participants.

The examples will describe in each case why we chose to use a database search. In general, we do a search if it allows us to answer a

89

question in a more timely and cost-effective manner, even though the information may be available by other means. The resulting savings in staff time have allowed us to handle a larger number of questions more efficiently. Second, we are finding that an increasing number of databases are providing information which only can be accessed in an online mode.

There is no participant fee for a simple MINITEX search. We have found that most questions do not require a great deal of online time. There is a charge for online and print costs for lengthier and more comprehensive searches; we sometimes suggest this course when we find it would augment the information we can provide for a request.

Because of the variety of questions received, the reference staff consider themselves subject generalists; they focus on what databases are available for what subjects rather than trying to become really proficient in searching individual databases. Searches requiring in-depth subject knowledge and experience with a particular database or range of databases are referred to MINITEX participants with that expertise.

I chose to use DIALOG databases for my search examples because it is the system with which I am the most familiar and, second, I felt that using one system would make for a less confusing demonstration. Applications of the kinds that I will be discussing, however, may also be made when using databases offered by SDC and BRS.

Development of a Bibliography

The first area of application I will illustrate is the development of a bibliography or list of sources on a particular topic. Because an online search of this type may make use of Boolean logic, multiple terms, and other features such as limitations by date and language, an online search can usually be done more quickly and comprehensively than a manual search. Second, access is available online to some indexes that in print would be too costly because of limited use.

In response to a request from a graduate student for information on camping programs for the emotionally disturbed, I searched three databases: ERIC, Exceptional Child, and PSYCHINFO. I have only illustrated the search on ERIC (figure 1) since the approach was similar on all three. I chose appropriate terms from the published list of ERIC descriptors and combined them using Boolean logic. Since the patron felt material published in the 1960s would be too old, I eliminated all citations previous to 1970. I have included in figure 1 two of the citations which were found, one an ERIC document and the second a journal article.

File 1* ERIC—66-82/Mar
Set Items Description
? s emotional disturbances or emotional problems
      2714 emotional disturbances
      634 emotional problems
  1 3303 emotional disturbances of emotional problems
? s resident camp programs or day-camp programs
      240 resident camp programs
      69 day camp programs
  2 289 resident camp programs or day-camp programs
? s py = 196?
  3 39343 py = 196?
? c(1 and 2) not 3
  4 11 (1 and 2) not 3
? t 4/3/1-2
4/3/1
ED204041
Evaluation of a residential therapeutic camping program for disturbed children.
Griffin, William H.
West Florida Univ., Pensacola. Educational Research and Development Center
20p. May 1981
EDRS Price—MF01/PC01 plus postage.
4/3/2
EJ190603
Wilderness camping: an evaluation of a residential treatment program for emotionally disturbed children.
Behar, Lenore; Stephens, David.
American Journal of Orthopsychiatry, v48 n4 p644–53 Oct 1978

FIGURE 1. ERIC search and two citations retrieved

The next example shows how I handled a request for recent medical literature on anorexia nervosa for a health paraprofessional. I performed this search on the MEDLINE file covering the time period 1980 to the present (figure 2). Anorexia nervosa is an established subject heading for MEDLINE, and when I entered it I came up with 374 citations. The problem then was how to limit this to a usable number. Since MEDLINE has both major and minor descriptors, I first limited my search to major descriptors. Second, I excluded all citations that were in a foreign language. Third, since review articles are tagged in the MEDLINE database, I chose to retrieve only review articles on anorexia nervosa, assuming the patron could get an overview and then pursue the topic further either using citations in the articles or by requesting a more specific search online. In figure 2 citations and abstracts are included for two of the review articles that were found.

```
File 154:MEDLINE—80-82/May
   Set Items Description
   ----------------------
? s anorexia nervosa
   1    374 anorexia nervosa
? limit 1/maj
   2    308 1/maj
? s la = english
   3    179220 la = english
? c 2 and c
   4    215 2 and 3
? s dt = review
   5    20678 dt = review
? c 4 and 5
   6    4 and 5
? t 6/7/1-2
6/7/1
0452315    81132315
```
Anorexia nervosa.
Schwabe AD; Lippe BM; Chang RJ; Pops MA; Yager J
Ann Intern Med. Mar 1981, 94(3) p371–81, ISSN 0003-4819 Journal Code: 5A6
The clinical and physiologic features of anorexia nervosa seem to be consequences of a complex interaction among psychologic abnormalities, endocrine disturbances, and malnutrition . . . (93 refs.)
6/7/2
0441584   81171584
Toward the understanding of anorexia nervosa as a disease entity.
Lucas AR

Mayo Clin Proc, Apr 1981, 56(4) p254–64, ISSN 0025-6196 Journal Code: LLY. Five major areas in the study of anorexia nervosa have been defined: the descriptive era (1868–1914), the pituitary era (1914–1940), the era of rediscovery of the illness (1930–1961), the psychoanalytic era (1940–1967); and the modern era (1961–). The illness as a biopsychological entity has been delineated. An integrated view of the disorder, recognizing multiple factor interaction in a vulnerable person, and an individualized approach to treatment, are proposed. (204 refs.)

FIGURE 2. MEDLINE search and examples of citations and abstracts retrieved

Nonbibliographic or Source Databases

The second area of application is the use of nonbibliographic or source databases. I have chosen several examples of how these databases provide actual business and statistical data online which eliminate the additional retrieval steps associated with bibliographic databases.

Requests for information on small companies are difficult because often nothing is available or easily accessible. However, recently a new database became available on DIALOG called Dun's Market Indicators 10+ which is compiled by Dun's Marketing Services, a company of the Dun & Bradstreet Corporation. This directory file, which covers over five million public and private U.S. companies having ten or more employees, is only available online and includes many businesses not covered in other sources. In figure 3 the location, marketing, and financial information that is available for Travelink, a travel agency with fourteen employees, is illustrated.

```
File 516:DUNS Market Identifiers 10+
(Copr. D & B) Mar 01 82
     Set Items Description
     ─────────────────────
? s Travelink
    1    1 Travelink
? t1/5/1
1/5/1
0664046
Travelink Trs Intnatl Inc
9575 W Higgins Rd
Des Plaines, IL 60018
Phone: 312-696-5790
   Cook County      SMSA: 118       (Chicago, Illinois)
   Business:                        Travel Agency
   Primary SIC:     4722            Passenger Transp Arrgm
   Year Started:    1973
   Sales:           $7,200,000
   Employees Here:  14
   Employees Total: 14
   This is:
   a single location
   a corporation
   DUNS number:                06-745-7937
   Corporate family DUNS:      06-745-7937    Travelink Trs Intnatl Inc
   Chief executive:            Rick Ricart Pres
```

FIGURE 3. Dun's Market Indicator 10+ retrieval on Travelink

My next example (figure 4) shows how I found statistical data for an individual preparing a talk on food consumption in the United States. This person wanted, among other things, a comparison of the consumption of butter and margarine over the past several years. I used the

```
File 16:PROMT—72-82/Apr
(Copr. PREDICASTS Inc.)
see file 17
      Set Items Description
      ----------------------
? s butter and margarine
         756 butter
         308 margarine
     1   85 butter and margarine
? s cc = 1USA
     2   265879    cc = 1USA    (cn = United States)
? s ec = 656
     3   10077     ec = 656     (en = consumption)
? C 1 and 2 and 3
     4   11 1*2*3
? t4/4/1-2
4/4/1
699951      J Amer Oil    81/11    p836a
US:         Visible fat consumption, 1980 (mil lb)
                                    per capita
butter                    1017         4.5
lard                       540         2.5
margarine                 2593        11.3
baking & frying fats      4158        18.3
salad & cooking oils      4820        21.2
other edible oils          343         1.5
total food               12722        55.9

4/4/2
582039      J Amer Oil    80/07    p538a
US per capita fats & oils consumption data (lbs)
                          1975      1978      1979
butter                     4.8       4.5       4.5
margarine                 11.2      11.4      11.6
lard                       3.0       2.2       2.3
shortening                17.3      18.2      19.2
other edible fats & oils  20.3      22.6      23.4
total                     53.4      55.6      57.6
```

FIGURE 4. Term search on PROMT database

PROMT database (Predicasts Overview of Markets and Technology) which abstracts relevant information from newspapers, business and trade journals, government reports, etc. Although this is published, I do not have access to the printed source and, in the print version, I could only search by subject. However, online I searched the two terms, butter and margarine, and because certain codes are assigned to each record. By using codes for the geographic location ("CC" stands for *country code*) and the type of statistical data I needed ("EC" stands for *event code*), I was able to retrieve information that only dealt with consumption in the United States. The PROMT database provides the statistical information online, so it is not usually necessary to retrieve the actual sources, many of which are in publications that are not widely held.

I tried to use searches that are simple and straightforward in order to illustrate the first two areas of application. However, depending upon the purpose, some searches are more involved and therefore more difficult to perform. In those cases, the searches are best performed by a person with relevant subject knowledge who has specialized in using databases related to that field and who has become familiar with the unique aspects of those databases.

Ready-Reference Applications

The third area encompasses a variety of searches which could be termed ready-reference applications. My examples include: verification, address or affiliation of author(s), current information, orientation to specialized terms or topics, and unique features of certain databases such as reviews and evaluations in the Magazine Index database, and acronyms and elusive publications in the Encyclopedia of Associations database. Here I am using a database search in order to quickly find some specific information which may not be accessible or may be difficult to find in print sources.

VERIFICATION

Complete bibliographic information was needed for an article by William Curtis on Le Corbusier's Villa Savoye, and some print sources, such as *Art Index*, had been checked. I decided to use the database Art-bibliographies Modern which includes books, periodicals, exhibition catalogs, and dissertations in the field of modern art and design. Although I have access to the printed version of this source, it is not located nearby, so I decided to do this quick search online. As illustrated in figure 5, the author was selected and combined with the architect's name and the name of his work. The article turned out to be in a book published in England.

> File 56:ART MODERN—1974-1981
> (Copr. ABC Clio Inc.)
> Set Items Description
> ? s au=curtis, w.
>   1   3   au=curtis, w.
> ? s corbusier
>   2   156   corbusier
> ? s villa(w)savoye
>   3   2 villa(w)savoye
> ? c 1 and 2 and 3
>   4   1   1 and 2 and 3
> ? t 4/7/1
> 4/7/1
> 071950
> History of architecture and design, 1890-1939, units 17-18: Le Corbusier; English architecture 1930s
> Curtis, W.
> Milton Keynes, England: Open University Press (1975), 104p. 104 illus. bibliog. Document type: book
> The ninth text in the Open University's integrated study course. Unit 17 analyzes the architectural vocabulary and meaning of Le Corbusier's 1929 Villa Savoye at Poissy . . .

FIGURE 5. Verification search on Artbibliographics Modern

ADDRESS OR AFFILIATION OF AUTHOR(S)

Some bibliographic databases include the address or affiliation of the author(s) in addition to citations to the literature. The following is an example of a request for this information: Who is doing research on cartilage hair hypoplasia, a rare growth disorder, and where are these research scientists located? I searched the term *cartilage hair hypoplasia* on two databases, SCISEARCH (Science Citation Index) and SSIE Current Research (Smithsonian Science Information Exchange). This was done as a database search because, since there is no subject approach to the database Science Citation Index, it was necessary to see if the terms appeared in titles, and SSIE is not available in print. Both SCISEARCH, which listed citations to current literature, and SSIE, which indicated current research, provided contact addresses online (figure 6).

File 34:SCISEARCH—81-82/wk10
(Copr. ISI Inc.)
See files 94 and 186
 Set Items Description

? s cartilage(w)hair(w)hypoplasia
 1   4 cartilage(w)hair(w)hypoplasia
? tl/3/1
1/3/1
0181882   OATS ORDER#:   lj987   23 refs
Cartilage-hair hypoplasia, defective T-cell function, and diamond-black-fan anemia in an Amish child (English)
Harris RE; Baehner RL; Gleiser S; Weaver DD; Hodes ME
Indiana Univ., Sch Med, Dept Med Genet/Indianapolis/IN/46204; Indiana Univ, sch Med, Dept Pediat, Div Pediat Hematol Oncol/Indianapolis/IN/46204
American Journal of Medical Genetics, v8, n3, p291–297, 1981
File 65:SSIE Current Research—78-82/Feb
(Copr. SSIE Inc.)
 Set Items Description

? s cartilage(w)hair(w)hypoplasia
 1   1 cartilage(w)hair(w)hypoplasia
? tl/3/1
3/3/1
0363669   SSIE NO.: 1GM 24736 4
Theoretical aspects of clinical genetics
Investigators: Murphy EA; Trojak JE; Abbey H; Rohde CA; Abbott MH
Performing Org: Johns Hopkins University, School of Medicine, Dept. of Medicine/Biostatistics, 725 N. Wolfe St., Baltimore, Maryland, 21205, United States of America
Sponsoring Org: U.S. Dept. of Health & Human Services, Public Health Service, National Inst. of Health, National Inst. of General Medical Science, 9000 Rockville Pike, Bethesda, Maryland 20205, United States of America
Contract/Grant No.: Rol GM 24736-04
1/78 to 12/81   FY: 81   Funds: $81,679

FIGURE 6. Author address/affiliation search on SCISEARCH and SSIE Current Research

CURRENT INFORMATION

Current information which cannot be found in print resources is often requested. For example, I received a request for up-to-date information

98 Management

on tar and nicotine levels in cigarettes. Assuming that this information would be issued in a government report, I first searched the GPO Monthly Catalog database by combining the terms *tar, nicotine,* and *cigarettes.* The latest report found, however, was for 1976. The same search was done for one newspaper on the National Newspaper Index database, the *New York Times* (figure 7). The National Newspaper Index is updated monthly, and the citations are listed in accession number order, with the most recent records given first. Since there were multiple terms involved, this information was retrieved more quickly through a database rather than a manual search.

Two citations retrieved are shown in figure 7. The first is to an article about the latest semi-annual report made by the Federal Trade Commission on tar and nicotine levels and includes the address of where to ob-

```
File 111: National Newspaper Index—79-82/Mar
(Copr. IAC)
     Set Items Description
     ----------------------
? s jn = New York Times
    1    264427      jn = New York Times
? s cigarette? and tar and nicotine
         359 cigarette?
          72 tar
          11 nicotine
    2      8 cigarette? and tar and nicotine
? c 1 and 2
    3    3 1 and 2
? t 3/3/1-2
3/3/1
0422698     DATABASE: NNI File 111
F.T.C. finds many cigarettes have less tar and nicotine.
New York Times v131 p13(n) pA23(LC) Dec 16 1981 CODEN:
   NYTIA
col 5     010 col in
EDITION: Wed
3/3/2
0160899     DATABASE: NNI File 111
Ranking the brands, (Cigarettes by amount of tar and nicotine)
New York Times v129 Section A pA22 Dec 17 1979 CODEN: NYTIA
col 3     004 col in
EDITION: Mon
Geographic Location: Washington
```

FIGURE 7. Current-information search on the National Newspaper Index database

File 32:METADEX—66-82/Apr
(Copr. Am. Soc. Metals)
　　Set Items Description
? s nitanol or nitinal or nitinol
　　　　0 nitanol
　　　　0 nitinal
　　　　30 nitinol
　　1　30 nitanol or nitinal or nitinol
? tl/3/1-2
1/3/1
856391　　81-110209
Energetic shape recovery associated with martensitic transformation in shape-memory alloys.
Golestaneh, AA
Acta Metall. Oct. 1980, 28 (10), 1427–1437
1/3/2
707501　　80-310912
Shape-memory alloys.
Schetky, LM
Sci. Am. Nov. 1979, 241 (5), 74–82
? tl/4/2
1/4/2
Shape-memory alloys.
. . . The alloy can be plastically deformed at one temp., and will completely recover its original shape on being heated to a higher temp. The transformation temp. of nitinol (NI-TI) . . . Unique applications for using these marmen (martensite-memory) alloys are given.—JJP
? B 47
File: 47:Magazine Index—77-82/Mar
(Copr. IAC)
　　Set Items Description
? s nitinol
　　1　4 nitinol
? tl/3/1
1/3/1
0577844　　DATABASE: MI File 47
Grassroots genius. (Suggestions of readers for using nitinol)
Sanders, Kevin
Science Digest v90 p26(4) March 1982 CODEN: scdibg
Illustration

FIGURE 8. Specialized term/topic search and METADEX and Magazine Index database

tain the complete report. The second is to an article which contains a chart of over 100 brands of cigarettes and their tar and nicotine content.

ORIENTATION TO SPECIALIZED TERMS OR TOPICS

I often receive requests for information on a topic with which I am unfamiliar and need additional background before I can search either print or nonprint sources. A representative example was the request for information about nitinol, "the metal with a memory," for a high school science fair project. I was also warned that the spelling might not be correct.

Since I had never heard of nitinol, I entered possible forms of the term on the database METADEX, as illustrated in figure 8. I was able to verify the spelling and obtain some good background information from the abstract of a *Scientific American* article on shape memory alloys. A second search was made on Magazine Index which provided less technical information.

    File 114:Encyclopedia of Associations—Edition 16
       Set Items Description
? s AAFDBI
    1   1 AAFDBI
? tl/5/1
1/5/1
00278
    American Association of Fitness Directors in Business and Industry
    (Athletic)   (AAFDBI)
    c/o Dennis Collacino, Ph.D., Pepsico, Anderson Hill Rd., Purchase, NY
      10577 (914) 253-2908
    Dennis Collacino Ph.D., Pres.
    Founded: 1974. Members: 1963. Regional Groups: 9. Exercise physiologists employed by major companies (some smaller companies and businesses are also represented) and conducting physical fitness programs for employees (529); interested persons in personnel and sales to fitness facilities, health educators (897); and students interested in this specialized field (539). Supports and assists in the development of quality programs of health and fitness in business and industry . . .
    Publications: Action (Newsletter), quarterly; also publishes membership directory. Proceedings of past conferences available on tape cassettes. Convention/meeting: annual. . . .

FIGURE 9. Unique-feature search on the Encyclopedia of Associations database

File 47: Magazine Index—77-82/Apr
(Copr. IAC)
    Set Items Description
? s Scarsdale(w)Diet
    1   14 Scarsdale(w)Diet
? s AT = Evaluation
    2   13292 AT = Evaluation
? c 1 and 2
    3   2 1 and 2
? t 3/5/1-2
3/5/1
0200025   DATABASE: MI File 47
Rating the diets. (Comparison)
Wise Woman's Diet.
Consumer Guide v223 p236(5) Spr 1979
ARTICLE TYPE: Evaluation
PRODUCT NAME(S): United States Senate Diet; Weight Watchers Diet . . . Scarsdale Diet; New You Diet; Dr. Stillman's 14 Day Shape-up Program . . .
DESCRIPTORS: Reducing Diets-Evaluation
4/5/2
0200005   DATABASE: MI File 47
Pushing Protein. (Comparison)
Consumer Guide v223 p103(25) Spr 1979
ARTICLE TYPE: Evaluation
PRODUCT NAME(S): Women Doctor's Womanly Diet; Pregnant Woman's Diet; Naturslim Diet; Hayden's Energy; Dr. Stillman's Quick Weight Loss Diet; Scarsdale Diet . . .
DESCRIPTORS: High-protein Diet-Evaluation; Low-Calorie Diet-Evaluation

FIGURE 10. Evaluation search on Magazine Index database

Unique Features of Particular Databases

Some databases contain unique information which can only be retrieved because of the ability to search them online. The first example (figure 9) shows my approach to a request for the full name and address of a physical fitness organization with the acronym AAFDBI, which could not be found in the *Acronyms, Initialisms, and Abbreviations Dictionary*. Acronyms, which are included in the printed *Encyclopedia of Associations*, are not indexed in the published version but are searchable online in the free-text mode. If you look near the end of figure 9 you will see that association publications are also included, so this is a good database to search for elusive publications which may not show up in regular serial and periodical sources.

My second example is illustrated by my approach to the question: Does the Scarsdale Diet work? The database Magazine Index tags all articles that provide consumer evaluations and then lists each product or service that is evaluated in an article so they are searchable. When I searched just using the words, *Scarsdale Diet*, I found 14 articles, but when that was combined with the tag, Article Type = Evaluation, I was able to limit my results to 2 articles which compared diets (figure 10). If this were done manually, with the microfilm edition of *Magazine Index*, it would be necessary to search under the Library of Congress subject headings relating to diets and dieting which are used by this index in order to find information on this specific diet. Once appropriate articles are located, it would then be necessary to examine all of them in order to discover evaluative articles only.

Ready-reference searches are usually done in a free-text mode and can be effectively performed by the generalist who knows the range of available databases and their scope. This final area is an excellent way for a library to introduce a database searching service and to provide experience to novice searchers.

GERTRUDE FOREMAN

The Policy Manual for Online Reference Services

During the past decade, library reference departments have experienced major changes brought about, in part, by the addition of computer-based reference services. The advent of online searching has challenged long-accepted reference philosophies and activities. Librarians have had to make decisions regarding management of and policies for online reference services. Too often these have remained informal or unwritten policies for providing services. Few libraries have a written policy manual for reference or online reference services.[1] But with computer-based services widely accepted, these formal, written policies are needed to record practice and to provide continuity. The decision to write a policy manual is a considerable commitment to a job taking time and intellect. It requires an exploration of (1) the need or purpose for policy statements, (2) the components of the policy manual, and (3) the possible approaches for policy formulation.

Definition of Policies

A policy may be defined as "a statement or general principle of library intent that helps translate program objectives into accomplishments by providing administrative guidelines for decision-making and implementation."[2] Policies are both restrictive and permissive, for they define the limits of acceptable practice and grant freedom for independent action within these limits. Furthermore, policy statements should address both major policies with library-wide implications and operating policies for day-to-day activities. Policy statements for online reference services fit within this framework.

Purpose of Policy Statements

The policy statement provides a basis for the management of the online reference service. It standardizes activities, facilitates decision making, and provides continuity. Specifically, the statements serve the following purposes:

1. To provide a framework directing activities toward a common goal
2. To establish performance goals and priorities
3. To assure consistency among client relationships
4. To protect the library's position should a particular staff action be challenged
5. To serve as an agreement between the reference librarian and the library administration.

To be useful, policy statements must be disseminated. The online reference policy manual is an appropriate means for communicating policies.

Components of the Policy Manual

No one policy manual can cover every question or decision, but the manual should include policy statements for all major aspects of online services. Every library, whether public, special, or academic, has distinct requirements and priorities. Although the policies differ among libraries, most online reference services share certain major areas of concern. The following are components to be considered for the online policy manual:

1. Statement on the purpose, goals, and objectives.
2. Statement on access and use policies. The statement should include clients served and the level of use provided. For example, the University of Houston manual clearly states not only patrons served but also groups not served.[3] Levels or priorities of service to clients should be defined. Do certain clients receive priority service?
3. Statement of service policies. A clearly articulated statement concerning online reference operations serves both the librarian and the client. Explain the scope of the service in terms of subject areas, specific databases, or selected information systems. If the library provides both for-fee and free searches, clearly stated guidelines are mandatory. When can the librarian perform a free reference search? Are demonstration searches and selective dissemination of information (SDI) searches available? How are requests accepted? By telephone or mail as well as in-person? State if an appointment is necessary and if the client must be present during the search. Describe turnaround time goals. Articulate those service policies which set standards of service. In addition to service policies, the manual may include details for handling

specific procedures. Responsibilities to be addressed might include the interview process, referral procedures, search results evaluation, and online search preparation.
4. Statement of financial policies. How are online services financed? Are there budget-related constraints? If charges are imposed, specify policies on user fees. Other financial policies should address delinquent accounts or refunds.
5. Statement of management policies. Clarification of personnel policies for online reference is essential. Who is in charge? To whom does this person report? Lines of authority should be clearly defined. What are the responsibilities of the charge librarian or coordinator? What is the responsibility of other reference librarians? How are they assigned and how is the work distributed? The statement should include a policy on continuing education for searchers.

Management concerns include policies on resolving client complaints, ensuring confidentiality of client requests, and coordinating online services within a larger system or network.

Preparing the Policy Manual

Development of the online policy manual depends upon the organizational structure and size of the library. In some libraries, one person may be responsible for preparing the manual, but usually a number of staff members are involved. A systematic approach facilitates the development process. Steps in the process may include: designing an overall plan; developing the list of areas requiring a policy statement; identifying and compiling existing policies; and formulating policies for approval and dissemination.[4]

Preparing the plan of action may involve only one person, but usually a task force or the total reference staff is involved in the development. This group can identify present formal and informal policies. Preparation of written versions of unwritten policies are probably necessary. Many policies evolve out of a series of decisions and have already been reflected in guidelines, fee schedules, staff minutes, and user brochures rather than in formal policy statements.

After compilation and preparation of existing policies, an evaluation is needed. What are the strengths and weaknesses of these policies? What are the problem areas? Are there gaps? Does the policy reflect the mission of the library? Is it consistent with other policies? What new policies need to be written?

Answers to these questions provide the framework for a comprehensive online manual. The manual should be sufficiently detailed to guide day-to-day activities while remaining flexible enough to accommodate changes. An appendix covering procedures and forms makes possible easy revisions for needed changes. A loose-leaf format is useful not only for expediting changes but for incorporating the online manual into the general reference manual, should this be desired.

The completed online manual must be approved or adopted by both the staff and administration. Policies which evolve from a shared planning process help achieve consensus on the goals for online searches. Thus the online reference manual becomes a powerful tool for improving online services and promoting needed changes.

Notes

1. Mary Jo Lynch, "Toward a Definition of Service: Academic Library Reference Policy Statements," *RQ* 11 (Spring 1972): 222–26; Mary E. Pensyl, "The Online Policy Manual: A Document That Few Have, but Many Can Benefit from (Especially Librarians)," *Online* 6 (May 1982): 46–49.

2. Duane E. Webster, "Library Policies: Analysis, Formulation and Use in Academic Institutions." Association of Research Libraries, University Library Management Studies Office. Occasional papers, no. 2 (Washington, D.C.: ARL, November 1972).

3. William J. Jackson, *Policy Manual for a Computerized Search Service in an Academic Library* (Texas: University of Houston, 1979, ED 174 221).

4. Webster, p. 11.

References

Association of Research Libraries. The Systems and Procedures Exchange Center. *Online Bibliographic Search Services.* Kit 76. Washington, D.C.: ARL Office of Management Studies, July 1981.

Atherton, Pauline, and Christian, Roger W. *Librarians and Online Services.* White Plains, N.Y.: Knowledge Industry Publications, Inc., 1977.

University of Massachusetts at Amherst. University Libraries. *Reference Service Manual.* Amherst: Massachusetts University, 1975, ED 116 701.

GAYLE McKINNEY

Forms and Record Keeping for Online Searching

A major area of search mechanics deals with forms and record keeping. Record keeping, one of the most time-consuming aspects of the search service, is also a vitally important one, necessary for both internal reporting and planning and for external auditing. In search services which charge a flat rate or fully subsidize their searches, the forms used may be simpler; nevertheless, most financial records are subject to audit and should be thorough and complete.

In this paper I will offer suggestions for record keeping in a new fee-based search service. Sample forms are provided and additional sources of information are noted. I will discuss types of forms needed, what information might be included on them, and how the information and forms could be used. Administrative concerns are also addressed.

Although some search services are now automating their records, a manual system is probably the most practical for a fledgling service and is the method I will discuss here. Conversion to an automated system can come later when needs are more thoroughly understood. In addition, automation requires some kind of computer capability and expertise which may not be readily available.

Development of Forms

Record-keeping forms should be developed so that the data collected will be complete and meaningful. Although sample forms are available from a multitude of sources, they must usually be adapted to a particular situation. Getting forms together, deciding on the ones necessary for

The author wishes to acknowledge with sincere gratitude the substantive comments by Mary Reichel and Jane Hobson of the Georgia State University Reference Department. Special thanks are due Margaret McMillan, GSU Online Assistant, for her dedicated editorial assistance.

complete record keeping, and making initial modifications is a time-consuming task in itself which should be done well in advance of the search service initiation. Practice searches and other internal searches should be recorded on the appropriate forms to see if any problems arise that require revision of the forms before the service officially begins.

What constitutes one search—each connection to the computer? each database searched? all databases searched for one question?—must be decided before selecting and preparing forms, because the answer affects some of the forms. A recent article by Hawkins and Brown gives various definitions for *search* and the advantages of each.[1]

Although additional modifications will take place as the service progresses, the forms should be basically in order when the first public search is offered, to prevent backlogs and gaps in the search service data.

Sources of Forms

Neighboring libraries with search services, particularly those libraries of a similar size and type, may be one of the best sources of sample forms. They have already done the work of developing the forms and they can point out any strengths and weaknesses in various examples. Also, these forms are readily available.

Books with good sections on forms include one by Wax and one by Atherton and Christian.[2] Comprehensive articles which include samples of forms are those by Hoover and Daniels.[3] The article by Hawkins and Brown, previously mentioned, contains a sample log sheet appropriate for the definitions they recommend. Finally, some libraries and organizations act as clearinghouses for forms. One of these, Machine-Assisted Reference Section of Reference and Adult Services Division, makes packets of collected forms available from ALA headquarters on loan.

Kinds of Forms

Two kinds of record-keeping forms are needed, those for use by the public and those for internal use.

PUBLIC RECORD-KEEPING FORMS

Forms used by the public include an application form, a payment/pickup form, an evaluation form, and a charge form.

The application or search request form (figure 1) includes information about the user including name, address, telephone, status, project deadline, and payment information. This form also requests information

```
Georgia State University
Reference Department    658-2185              Appt. Date _____ Hour _____

REQUEST FOR A COMPUTERIZED LITERATURE SEARCH   Project dealine date _____

Person for whom the search is to be made:     Maximum willing to pay
                                                 for this search_____
_____ ( ) New User
Name (Last name first - please print) ( ) Previous User    PAYMENT:
                                                           Check/money order___
                                      ( ) GSU Faculty
_____        ( ) GSU Staff            Dept.chg./grant ___
Department                            ( ) GSU Ph.D. Candidate
                                      ( ) GSU Master's Candidate   TELEPHONE:
_____        ( ) GSU Undergraduate
On campus address                                              Home_____


_____        _____
Name of person requesting the search  University ID number    Work_____

The online search process usually requires about ten (10) working days for
completion and the GIDC off-line process usually requires 2-3 weeks for completion.
Please allow ample time. While searches are formulated and conducted with care, no
guarantee of completeness or accuracy can be made. Submission of this form commits
the requester to the payment of charges incurred. The Departmental charge card or
grant charge card should be presented prior to the execution of the search; for
those paying by check, a personal check will be required before the search results
are picked up.

SEARCH INTEREST Please write a sentence or paragraph which clearly describes your
search topic. Be as specific as possible and define terms which have a special
meaning.




DESCRIPTORS/KEYWORDS Please list descriptors, keywords, synonyms, acronyms, index
terms, or phrases frequently used to describe this search topic. Group related
terms according to the major ideas in your question. Attach an additional sheet if
necessary.




RELEVANT PAPERS List several papers on your topic which you consider key articles.
Please give complete citations. These titles help the search analyst to understand
the topic and frequently provide important clues for additional keywords and search
strategy. Attach an additional sheet if necessary.
```

FIGURE 1. Search request form

on the search topic and scope of the project. Requiring that the search topic be written out in sentence or paragraph form can be of primary importance in helping the searcher understand the topic. Another important item on the form is the listing of keywords or thesaurus terms.

110   Management

>
> AUTHORS If there are particular authors whose writing is _always_ of interest to you in relation to this topic, please list them, giving at least the last name and first initial.
>
> PREPARATORY SEARCHING List the indexes you have already searched for articles on this topic.
>
> COVERAGE If you know which data bases you want searched, please list them.
>
> LIMITATIONS List below any relevant descriptors or keywords describing concepts for ideas to be _excluded_ from this search.
>
> TIME COVERAGE Complete retrospective coverage_____Other(specify years)_____
>
> LANGUAGES English only_____ Languages in addition to English_____
>   Any languages_____
>
> SCOPE References needed for: Research proposal_____ State of the art review_____
>   Paper/publication_____ Book_____ Talk_____ Dissertation_____ Thesis_____
>   Term paper_____ Class assignment_____ Other(specify)_____
>
> SEARCH REQUIREMENTS Broad or Narrow
>
> ( ) A BROAD search designed to retrieve as many citations relevant to your topic as possible, but which might also retrieve many citations not relevant to your topic.
>
> ( ) A NARROW search designed to retrieve citations directly relevant to your topic, but which might not retrieve all relevant citations.
>
> NUMBER OF CITATIONS Estimate the number of journal articles published on your topic within the past twelve (12) months_____. What is the maximum number of citations you want from this search?_____
>
>   I request that the library make this search on my behalf and I accept the responsibility for paying for the search
>
>   _____
>   Signature   Date
>
>   When you have completed this form, please return it to the Reference Department. Your search request will be studied by a Reference Librarian who will arrange an appointment to discuss the details of your search.

FIGURE 1. Continued

Having the keywords listed in advance by the user can frequently shorten the search interview. The listing of one or two pertinent articles which can be checked for descriptive terms is also valuable.

Depending on their proposed use, search application forms may be simple, requesting only a few key items, or they may be quite comprehensive. A fairly detailed form, such as the sample, can guide the user in thinking through the search topic and project scope and remind the searcher to check on all the necessary points. Fenichel and Hogan give a sample of a very brief search request form and list additional items to be included in a more comprehensive version.[4] They also state that a more elaborate form can include space for recording search results, such as databases searched and costs. Finally, unless the service is free, a place should be provided on the form for the user to sign, agreeing to pay for the search.

The search request is used in other record-keeping activities and should be retained with a record of the search for a reasonable length of time. Two years is sometimes specified. The forms, the search record, and other material relevant to the search may be kept in a folder under the user's name. Questions regarding the search and possible expansions or updates of it may arise later. Financial questions may also develop.

A payment/pickup form (figure 2) is a useful aid for library staff who distribute search results to patrons and receive payments. The pickup form can be retained, too, as a permanent record of the search. The form is filled out using information from the search request and the log sheet, and may be filed in the user's folder until the search results arrive. It is then attached to the prints. The form should have enough information (name and telephone number) to allow the patron to be notified. Search information such as log number, databases searched, transaction dates, and charges is also recorded. If required by the institution, the user's financial status can be checked in advance and the payment method specified accordingly. Users should sign and date the form when they receive the prints and the staff member handling the transaction should initial the form and enter the user's check or charge account number. A receipt book should be kept handy, since occasionally a user will ask for a receipt, even if paying by check. The payment/pickup forms can be kept with the checks or charges until these payments are forwarded to the business office, and then kept with an adding machine tape or other listing of the payments as a permanent record of the deposit and a concise record of the searches.

If internal fund transfers are to be made for searches charged to departments or grants, a multipart carbon charge form (figure 3) will be needed. A form already in use by the institution for charging other services can possibly be adapted. The form should have at least three parts: one for the user, one for the search service records, and one for the business office to attach to long-term grant records. At this time, federal regulations require that financial records connected with federal grants be kept for five years after the completion of a project.

```
         Georgia State University          Phone Numbers
         Reference Department
                                       (H)_____
         COMPUTERIZED LITERATURE SEARCH
                   CHECKLIST            (W)_____

       Name_____
                  (last)                (First)

       UNIVERSITY ID #  _____

         ┌─────────────────────────────────────────────┐
         │  FIN. HOLD CHECK   BY____  Status   OK ___  │
         │                            Money Order Only __│
         └─────────────────────────────────────────────┘

       LOG SEARCH #  _____

       DATE BASES SEARCHED  _____

       TRANSACTION DATES

           Request rec'd_____  Search run _____ By_____

           Output rec'd _____  Evaluated by _____

           Patron notified by _____ on _____

               via phone _____  mail _____

         ┌─────────────────────────────────────────────┐
         │       Total charges:      CASH NOT ACCEPTED:│
         │                           Make check payable│
         │     Library pymt.(-)____  to GSU Library.   │
         │                           Checks must show  │
         │   REQUESTER CHARGES:      GSU ID#.          │
         └─────────────────────────────────────────────┘

       Paid by:  Personal check# _____

       Contract/Grant/Dept.Acct.# _____

       Picked up by _____ Date ____ Librn____
```

FIGURE 2. Payment/pickup form

At the end of the month, a summary list of departmental and grant charges, with copies of the multipart forms attached, should be sent to the business office to be charged to the proper accounts. A summary sheet for listing such charges is usually available from this same office.

WILLIAM RUSSELL PULLEN LIBRARY
Charge Account Record

☐ Computer Search
☐ Interlibrary Loan
☐ _____ Other

TOTAL _____

Description of transaction:

40-49

Signature _____

Date _____

FIGURE 3. Charge form

In corporate settings, the forms and procedures for internal fund transfers may be somewhat different. Typically, no charge forms are used. At the end of the month, the coordinator compiles a summary list of searches by department or account numbers and forwards the list and the vendor invoice to the accounting department for payment. In many cases, however, corporate library budgets absorb the entire cost of online searching and no internal fund transfers occur.

An evaluation form (figure 4) may be given to the user with the search results to be returned after the results have been examined. The answers to the questions usually provide marketing information (e.g., Who is using the service? What for? How did they find out about it?), as well as an evaluation of the databases used and the search process.

INTERNAL RECORD-KEEPING FORMS

Internal record-keeping forms include the following: scheduling sheets; log sheets, monthly activities sheets; and monthly statistical forms.

The method of scheduling varies from library to library but, in general, a scheduling form (figure 5), broken-down by hours for at least a week ahead, is desirable. An appointment book which has hour or half-hour divisions may also be used. The scheduling form needs to have room for the user's name and telephone number (in case the appointment needs to be rescheduled), the searcher's initials, and the databases or subject field to be searched.

A log sheet (figure 6) contains a record of each search or computer connection, with user information taken from the request form and search information taken from the terminal output. Each search is also assigned a log number and has the searcher's initials recorded. A notes column (not shown on the sample) is extremely helpful for recording miscellaneous information such as stored search numbers and trouble on the system. The log is to be used in compiling statistical reports and in verifying charges on the monthly vendor bills. To avoid having a misplaced check go unnoticed, the log should have a column showing the date each user's payment is deposited. Many institutions prefer to keep a separate log for each vendor; this is especially desirable if the billing formats differ greatly. The separate logs can then have different formats to facilitate the checking of each vendor's bills. If bills are arranged in chronological order by password, a separate log may be needed for each password. Free searches offered as part of regular reference service may also require their own log, since libraries frequently want to keep a special record of these.

EVALUATION OF ONLINE COMPUTER SEARCH

Name _____ Date _____ Database _____ ☐ Original ☐ Updates

1. Type of User: Faculty _____ Staff _____ Graduate Student _____ Undergraduate _____ Other _____

2. Where did you learn about the Computer Search Service?

   Reference Desk _____ Library Orientation Class _____ Printed Announcement _____

   Friend/Colleague _____ Professor/Adviser _____ Other _____

3. What accounts for use?

   Thesis/Dissertation _____ Class assignment/Term paper _____ Publication/Project _____

   Other research _____ Other (Please explain) _____

4. Previous search experience: Is this your first search? Yes _____ No _____ If no, in how many have you participated? _____

5. Evaluation Statements:

	Strongly Agree	Agree	Undecided	Disagree	Strongly Disagree	Not Applicable
a. Useful information was obtained from this search.						
b. The database was adequate in its coverage.						
c. Interaction between the user and the librarian contributed to the success of searching.						
d. The printout format was satisfactory.						
e. I consider the services to be generally cost effective.						
f. I would use this service again.						

6. Comments and suggestions:

Note: Please return this self-addressed evaluation form through the general campus mail or bring it by the library Reference Department after you have received your complete set of retrospective printouts.

FIGURE 4. Evaluation form

116 Management

Computer Search Schedule		Month:	Year:	
Monday	Tuesday	Wednesday	Thursday	Friday
8:00	8:00	8:00	8:00	8:00
9:00	9:00	9:00	9:00	9:00
10:00	10:00	10:00	10:00	10:00
11:00	11:00	11:00	11:00	11:00
12:00	12:00	12:00	12:00	12:00
1:00	1:00	1:00	1:00	1:00
2:00	2:00	2:00	2:00	2:00
3:00	3:00	3:00	3:00	3:00
4:00	4:00	4:00	4:00	4:00
5:00	5:00	5:00	5:00	5:00
6:00	6:00	6:00	6:00	6:00

FIGURE 5. Scheduling form

ONLINE COMPUTER SEARCH LOG																			
			USER INFORMATION										USER PYMT INFO			INVOICE INFORMATION			
LIBN	DATE	SEARCH #	NAME	DEPT	TYPE	SYST NAME	FILE #	START TIME	END TIME	ESTM COST	# TYPES	# PRNTS	PATRON CHARGE	CK#/ ACCT#	DEP DATE	INVOICE #	DATE	AMT BILLED	CLRD

FIGURE 6. Log sheet

118 Management

A monthly computer-search activities sheet (figure 7) can serve as a calendar of upcoming events such as training sessions, demonstrations, and meetings. It can also be used to record search-service milestones such as the receipt of a new high-speed terminal. This type of nonstatistical information provides highlights for monthly and annual reports.

COMPUTER SEARCH ACTIVITIES		Month:	Year:	
Monday	Tuesday	Wednesday	Thursday	Friday

FIGURE 7. Computer-search activities sheet

Gayle McKinney 119

Without such a calendar, nonstatistical information is easily lost or forgotten.

The final samples shown are a monthly statistical summary sheet (figure 8a) and a supporting detail sheet broken-down by database (figure 8b). These statistics are taken from the log sheets and the monthly

```
            ONLINE SEARCH SERVICE STATISTICAL REPORT
                    MONTHLY SUMMARY FOR _____
```

	CONNECTS*		REASONS FOR CONNECTIONS (Current Month)				
VENDORS	Year to date	Current Month	Fac/ Staff	Students G / U	Off Campus	TOTAL PATRON	TOTAL OTHER
DIALOG							
MEDLARS							
TOTALS							

			GAIN/(LOSS)	
FINANCES	Monthly Cost	Monthly Revenue	Current Month	Year to date
DIALOG				
MEDLARS				
MISC CHGS				
SUBTOTALS				
CREDITS				
TOTALS				

TOP 5 DATABASES:

TOP 5 DEPARTMENTS:

PATRON SEARCHES**	Year to date	Current Month
NUMBER		
COST: High		
Low		
Avg.		
DIALORDERS		
CON.HRS.USED: DIALOG		
MEDLARS		

* CONNECTS - All database searches for any reason including "patron" searches (faculty/staff, student, off-campus) and "other" searches (demonstrations, practice, miscellaneous, and library department).

** PATRON SEARCHES - Database searches done for faculty, staff, students, and active alumni.

SIGNIFICANT EVENTS:

FIGURE 8a. Monthly statistical summary sheet (first example)

120 Management

DIALOG	Fac/Staff	Students G	Students U	Off-Campus	TOTAL PATRON	Total Other	Current Month	Year to date
1 ERIC								
2 CA SEARCH 67-71								
3 CA SEARCH 72-76								
4 CA SEARCH 80-81								
5 BIOSIS PREVIEWS								
6 NTIS								
7 SOCIAL SCI SEARCH								
8 COMPENDEX								
9 AIM/ARM								
10 AGRICOLA								
11 PSYCINFO								
13 INSPEC 77-								
15 ABI/INFORM								
16 PTS PROMT								
17 PTS PREDALERT								
18 PTS F&S INDEX 79-								
19 CHEM INDUSTRY NOTES								
22 EIS INDUST PLANTS								
27 FOUNDATION GRANTS								
31 CHEMNAME								
34 SCISEARCH 81-								
35 COMPRE DISSER IDX								
37 SOCIOLOGICAL ABS								
38 AMERICA: HIST & LIFE								
39 HISTORICAL ABS								
40 ENVIROLINE								
46 NICEM								
47 MAGAZINE INDEX								
49 PAIS INTERNATIONAL								
51 FOOD SCI & TECH ABS								
52 TSCA INITIAL INVENTORY								
SUBTOTAL								

ONLINE SEARCH SERVICE STATISTICAL REPORT FOR _____

VENDOR/FILE — CONNECT CATEGORIES — TOTAL CONNECTS

FIGURE 8b. Monthly statistical summary sheet (second example)

vendor invoices. Compiling key statistics on a monthly basis is very helpful for a new search service as well as for an established one. Statistics reassure both you and your administration that the venture is going well and provide information for future planning. The types of statistics compiled depend on what the service wishes to emphasize, but usually include the number of searches by type of user and by type of search. Money spent and collected is reported, as well as the number of connect hours used. In colleges and universities, a breakdown by departments using the service may be valuable.

Statistics for the monthly detail sheet (figure 8b) are fairly time-consuming to compile manually, and if individual database totals are not needed, a new service might prefer to compile only those statistics required on a brief form such as the summary sheet (figure 8a). Another short monthly statistics form with several sections taken from the vendor invoice is shown on page 39 of Hoover's article.

Administrative Concerns

Extensive record keeping is time-consuming and can easily become quite costly in staff hours. Recording only what is really necessary and/or useful in your situation helps keep the cost under control. Assigning the record keeping to a clerical assistant rather than a searcher is also advisable. The assistant should be carefully chosen. Although the majority of the work is fairly routine once the forms are set up, the record keeping requires mathematical aptitude and the ability to handle a great deal of detail efficiently. An alert assistant can also advise on possible forms revision once the service is underway.

Before deciding on the final formats of the initial forms, the business office of the institution should be consulted to see what information it requires. If a publication or printing department is available, it can be consulted as to the final design and actual production of the forms.

Conclusion

Well-designed record-keeping forms can make reports and statistics simpler to prepare and can help keep your service in good standing with the business office and the auditors. These forms can also enhance the presentation of your accomplishments to the administration, which will insure the continuation and expansion of the service.

Notes

1. Donald T. Hawkins and Carolyn P. Brown, "What Is an Online Search?" *Online* 4, no. 1 (January 1980): 12–18.

2. David M. Wax, *A Handbook for the Introduction of Online Bibliographic Search Services into Academic Libraries*, Office of University Library Management Studies. Occasional papers, no. 4 (Washington, D.C.: Association of Research Li-

braries, 1976); Pauline Atherton and Roger W. Christian, *Librarians and Online Services* (White Plains, N.Y.: Knowledge Industry Publications, 1977).

3. Ryan E. Hoover, "Computer Aided Reference Services in the Academic Library: Experiences in Organizing and Operating an Online Reference Service," *Online* 3, no. 4 (October 1979): 28–40. Linda Daniels, "A Matter of Form," *Online* 2, no. 4 (October 1978): 31–39.

4. Carol H. Fenichel and Thomas H. Hogan, *Online Searching: A Primer* (Marlton, N.J.: Learned Information, 1981).

NANCY E. GRIMES

Costs, Budgets, and Financial Management

An online reference service entails costs above and beyond those of traditional nonautomated reference services. Online searching requires special professional skills, extra administrative policies, and new financial considerations which include, among other things, the cost of computer terminals and the decision of whether or not to charge fees. An important component of an online search service is a creative, flexible, and relevant budget. The costs of online searching must be investigated and an initial budget created which reflects the scope and goals of the service to be provided. Preferably, such a budget is designed as part of an initial proposal to introduce online capabilities into the reference environment; certainly a budget should be established before any steps are taken to initiate service. Annual or semi-annual reviews, supported by the statistics of experience, produce ongoing budgets which can promote controlled management of the service within the library or information center.[1]

In this paper, the cost elements of establishing and running an online reference service will be outlined, and initial and ongoing budget structures suggested, with some hypothetical examples for consideration. Enough elements are involved in the financial aspects of online reference to provoke a brief discussion of cost management and the options available, which range from the type of equipment acquired to discount contracts to possible methods of funding. Although the last topic is outside the scope of this paper, funding methods—including whether or not to charge patrons for the service—must be coordinated with cost accounting and budgeting.

Cost Elements

The costs involved in online searching can be conveniently divided into direct and indirect categories, as shown in figure 1. Direct costs are,

I. Direct
   A. Search sessions
      1. Connect time
      2. Telecommunications
      3. Citation charges
      4. Labor: searcher's time
II. Indirect
   A. Acquisitions
      1. Equipment
         a) terminal
         b) modem
         c) telephone
         d) slave printer
         e) maintenance contract(s)
      2. Documentation and supplies
         a) terminal paper, ribbons
         b) manuals
         c) thesauri
         d) subscriptions
         e) professional memberships
   B. Operating expenses
      1. Personnel
         a) coordinator/administrator
         b) librarian/searcher(s)
         c) support staff
      2. Training expenses
         a) vendor training
         b) database, subject-oriented seminars
         c) practice time online
         d) travel expenses
      3. Promotion
         a) printing and graphics
         b) online time for demos, complementary searches
      4. Overhead
         a) facility use and modification
         b) furniture
         c) utilities

FIGURE 1. Cost elements

quite simply, the costs incurred by an online search session: the connect-time charge for using a vendor and database, the charge for the use of a telecommunications network or other telephone line, and the charges for citations to be printed offline. Since some databases also charge separately for each citation viewed online at the terminal, the budget entry reads "citation charges," encompassing both online and offline charges. The time spent by the searcher preparing and performing the search may be calculated and itemized under direct costs; alternatively, it may be considered to be part of the indirect cost of personnel.

Indirect costs include all other expenses involved in running an online reference service and can be categorized as acquisitions and operating (or recurring) expenditures. Acquisitions covers direct purchases to be made, here divided into equipment, documentation, and supplies. The necessary equipment components are terminal and modem (or a combined unit) and a telephone line, preferably dedicated exclusively to online activities. This equipment may be leased or purchased. Maintenance contracts for equipment should also be listed as a budget item. Optional pieces of peripheral equipment, such as floppy-disc storage units, would be listed here.

Documentation and supplies must also be purchased. If equipment includes a printer, paper will be needed; if the printer is an impact type, new ribbons will be needed. System manuals, database thesauri, and other user aids will compose the search service's own reference collection, together with system newsletters (usually free to system users) and subscriptions to relevant journals. The professional staff may wish to join an online user group or participate in online-related activities in a national organization; membership and conference fees should be considered as a budget item if the institution provides the funds.

A variety of recurring cost elements constitute the category of operating expenses, but this heading does not imply that the costs categorized as "direct" and "acquisitions" do not recur. The indirect recurring cost elements cover four general areas—personnel, training, promotion, and overhead. The specific line items for these elements may vary. For example, searcher and coordinator may be one and the same person, or coordination may consume a quarter of a department head's time. Training sessions may take place on site or may require travel and overnight lodging. Promotion may include newsletters, brochures, or complementary searches. The search-service facility may need modification, and so on. The cost elements listed here may serve as guidelines for budget configuration.

Initial Budget

Appropriate cost elements are put into an initial, or start-up, budget as illustrated in figure 2. Preparatory costs, listed first, are those incurred in the activities which take place before the first search is even run. Equipment, supplies, and basic manuals and thesauri must be ac-

I. Preparatory costs
   A. Equipment
      1. Terminal and modern purchase/lease
      2. Telephone installation
      3. Maintenance contract(s)
   B. Facility
      1. Modification
      2. Furniture
      3. Utilities
   C. Documentation and supplies
      1. Terminal paper, ribbons
      2. Manuals and thesauri
      3. Subscriptions
   D. Search system contract and network membership

II. Operating expenses
   A. Personnel/labor
      1. Coordinator/administrator
      2. Librarian/searcher(s)
      3. Support staff
   B. Training expenses
      1. Vendor training
      2. Practice time online
      3. Travel expenses
   C. Promotion
      1. Printing and graphics
      2. Online demos, complementary searches
   D. Search sessions
      1. Connect time
      2. Telecommunications
      3. Citation charges
      4. Labor

FIGURE 2. Initial budget

quired. The facility must be made ready, which may involve possible remodeling and furniture purchases. Search-system (vendor) contracts must be established, and while the contracts usually involve no start-up fee, under some arrangements advance subscription fees are paid, or a library may decide to join a library network that requires a membership fee in order to take advantage of a group discount. These options will be discussed further within the theme of cost management. Among operating expenses, personnel time will be required to set up the service and attend training sessions. Expenses for those sessions must also be included. Introductory system training frequently provides for limited free practice time at home, but it may be useful to allocate funds from the budget for additional online practice time. Promotion pieces—flyers, brochures, posters, or other announcements—will be produced, and introductory demonstrations prepared.

An initial budget could be extended to include the first three, six, or even twelve months of running the service.[2] In this case the direct costs of the search sessions must be included in the budget. These are costs difficult to estimate in advance; the prices of the search systems and databases used will affect the expense, as will the amount of searching done and the fees which may be charged. There are methods for arriving at an estimated cost for these items (connect time, telecommunications, citation charges, labor). For instance, the scope of the service to be provided could be limited to only one or a few low-cost databases on one system, or to a defined clientele group. Colleagues already searching in a service whose scope and clientele are similar can be consulted about their cost experiences. A policy charging the client all search session costs can be implemented and the figures excluded from the budget. The literature can be consulted for further suggestions and case studies; some citations will be found in the bibliography.

Operating Budget

Once an online reference service has opened, the initial budget may be expanded and adapted to reflect the ongoing expenses of providing searches. The preparation of this budget will take into account the cost statistics kept during the previous months, utilizing recent past experience. The budget will also reflect the plans for the future of the service, possibly including cost estimates for purchase of new equipment, addition of staff, improvement of the facility, or increased search activity.

The recurring expenses of running the service, as well as anticipated costs of expansion, are reflected in the budget items outlined in figure 3. Following the same general pattern of the initial budget, some additional cost elements now appear. Equipment costs may include new purchases but primarily reflect maintenance expenses. Supplies are, of course, an ongoing expense; additional manuals and thesauri, as well as new editions of old documentation, constitute another. Professional association expenses are incurred as staff participate in meetings and other activities. Network memberships may have to be renewed.

Personnel costs are continuous and may increase as salaries go up or new staff is added. Training should be viewed as an ongoing activity: searchers need advanced and specialized training, and new staff require

I. Indirect costs
    A. Equipment
        1. Terminal and modem—maintenance and upgrading
        2. Telephone use
        3. Maintenance contract(s)
    B. Documentation and supplies
        1. Terminal paper, ribbons
        2. Manuals and thesauri
        3. Subscriptions
        4. Professional memberships and meetings
    C. Network membership
    D. Personnel/labor
        1. Coordinator/administrator
        2. Librarian/searcher(s)
        3. Support staff
    E. Training expenses
        1. Advanced vendor training
        2. Database and subject-oriented seminars and workshops
        3. New staff training
        4. Practice time online
        5. Travel expenses
    F. Promotion
        1. Printing and graphics
        2. Online demos, complementary searches
    G. Overhead
        1. Facility use and modification
        2. Furniture
        3. Utilities
II. Direct costs
    A. Search sessions
        1. Connect time
        2. Telecommunications
        3. Citation charges
        4. Labor

FIGURE 3. Operating budget

introductory preparation. Continuing promotion will involve new brochures, demonstrations, etc. Overhead costs may seem to consist primarily of utility bills, but use or modification of space and the purchase of furniture may now be included in this category rather than under preparatory expenses.

The cost estimates of the search sessions themselves may now be based partly on experience, provided that forseeable changes are accounted for. Systems generally announce price increases in advance; inflation dictates that there be some regular increase in the basic prices to be paid. Search costs also increase as the scope of the service provided expands and as search volume increases. On the other hand, as volume increases, price discounts may be sought through system contracts. The same methods for estimating these costs mentioned earlier may again be resources for budget calculations.

Cost Management

Some estimated cost figures are offered in two sample budgets (figure 4) intended to cover preparatory costs and operating expenses for three months. The Machine-Assisted Reference Section (MARS), Costs and Financing Committee, compiled the cost information on which these figures are based; that committee may be contacted for details about the figures quoted.

The scope of service A can be described as modest and relatively low-cost, projecting forty searches in the three-month period, or three to four per week, in a limited group of databases on one system. Service B is more ambitious, has more capital, and plans eighty searches on one system in the period defined.

Search service A purchases a 300 baud thermal printing terminal with acoustic coupler and installs a phone and an electrical outlet. No other modifications are made to the facility. Documentation and supplies include three rolls of terminal paper, a search system manual, ten system database descriptions, and one journal subscription. Personnel costs of service A are based on the estimated labor hours required of one professional searcher-coordinator and one paraprofessional support staff member to provide the defined level of service. Neither staff member is dedicated full-time to the service. The searcher-coordinator attends an introductory system training session, incurring some travel expense, and is budgeted $50 for online practice to supplement the training course. Promotion takes the form of a locally produced mimeographed flyer and five strategically planned demonstrations. Search session costs are estimated for forty twenty-minute searches at an average connect-hour price of $50, with telecommunications at $7 per hour, and thirty citations per search at an average 12¢ each. Total initial budget: $4186.

Service B purchases a 1200 baud CRT with attached impact printer and separate modem. A phone is installed, and $2000 budgeted to modify an alcove into a search station. Along with terminal paper and inked ribbons, funds are provided to purchase a system manual, thirty data-

Preparatory Costs		A	B
Equipment			
terminal and modem		$1695	$4635
telephone		138	138
maintenance contract(s)		60	228
	TOTAL	$1893	$5001
Facility			
modification		$ 55	$1400
furniture		0	500
utilities		45	100
	TOTAL	$ 100	$2000
Documentation, supplies			
paper, ribbons		$ 9	$ 30
manuals and thesauri		48	430
subscriptions		78	134
	TOTAL	$ 135	$ 594
Operating Expenses			
Personnel/labor			
coordinator/administrator		$ 550	$ 620
librarian/searcher(s)			688
support staff		120	306
	TOTAL	$ 610	$1614
Training expenses			
system training		$ 35	$ 500
practice time		50	100
travel expenses		100	0
	TOTAL	$ 185	$ 600
Promotion			
printing, graphics		$ 40	$ 400
demos, free searches		260	520
	TOTAL	$ 300	$ 920
Search sessions			
connect time		$ 666	$1600
telecommunications		93	160
citation charges		144	288
	TOTAL	$ 903	$2048
Total Initial Budget		$ 4,186	$12,777

FIGURE 4. Sample initial budgets

base descriptions, and seven primary thesauri. Two journal subscriptions are established. The personnel budget of service B covers the labor hours of one coordinator, four searchers, and support staff member, again none dedicated full-time. Introductory system training, including online practice, is held on site; funds for additional practice time are also set aside. Promotion involves a printed brochure and ten online demonstrations. Though service B plans to access only one search system, no database restrictions are planned for the six to seven projected searches per week. The average connect cost used here is slightly higher at $60 per hour. Total initial budget: $12,777.

The cost estimates provided here are realistic, but the budgets should be viewed as exemplary rather than definitive. Not only do actual dollar figures very, but the cost elements to be included will depend on a multitude of decisions abut the scope and structure of the services to be provided. Additionally, the budget itself can be arranged in any number of ways. Many other models can be found in published articles and case studies.

Two areas of major expenditures—equipment and search system, or vendor, contracts—deserve special mention because of the broad range of cost strategies available. In the area of equipment, computer terminal purchase price alone can vary from the basic combined unit at $1695 included above in budget A to the $4695 spent in budget B for a higher speed, more flexible set of components to $5000 or more for a personal microcomputer with storage to $12,000 or more for a modular, upgradable microcomputer.[3] Additionally, equipment can be leased rather than purchased, an option sometimes attractive for a service just getting started or in the experimental stage. Other resources for equipment may include a campus computer department, a wealthy patron, or a library-friends group.

Search system contract options vary widely. While there are similarities among those offered by the major vendors, there is no standard, and each option for each vendor must be studied with care. The data here is accurate as of June 1982, but can change rapidly. The best source for current information is the vendor itself. None of the major systems charge a start-up fee. The simplest contract option, which is available from DIALOG, SDC, and NYTIS, requires neither prepayment nor usage guarantee and involves simply a monthly charge for the connect time and telecommunications time and citation charges. BRS now also offers a pay-as-you-go option, with a $50 prepayment toward connect time.

Some systems, including DIALOG and SDC, automatically discount all accounts after a certain level of usage each month—for example, after the first five hours connect time a $5 per hour discount begins to be applied—and the discount increases as usage increases. Most vendors, however, also offer higher discount possibilities in exchange for prepayment or a usage guarantee, options generally described as subscriptions and group plans. Contract details vary from vendor to vendor. Subscriptions may be established for either single passwords or groups of passwords on one account. The common element is an annual usage guarantee.

The standard individual contract with BRS requires a subscription in the sense of guaranteed annual usage; prepayment is preferred but not required, and group plans are encouraged. The higher the level of usage subscribed for, the lower the connect-hour rate. DIALOG's subscription option requires at least two passwords per account and prepayment is required; connect-hour discounts are received in return. DIALOG also offers an individual monthly usage guarantee contract with connect-hour discounts and no prepayment.

The New York Times block pricing plan involves an annual usage guarantee at one of seven levels, with monthly invoicing and no prepayment. The connect-hour rate depends upon the level of usage guaranteed, and more than one password may participate in a single guarantee.

SDC's automatic discount mentioned above is provided at the account level, and one account may have multiple passwords, which can belong to different users, thus creating a group plan. Neither a guarantee nor prepayment is involved.

Most group plans offer discounts in exchange for guaranteed usage on either a monthly or an annual basis. The advantage of any group contract lies in the protection that the group affords in meeting the level of usage guaranteed, and in the group's capacity to guarantee at a higher level than any one user would be able to do. Both BRS and DIALOG provide this option. While it is certainly possible simply to form a new group, there are also groups already in existence which are open to new participants. The best known of these are operated by library network organizations such as AMIGOS, BCR, CLASS, FLC, and NELINET. Participants can frequently choose from more than one contract option and can design a contract that does or does not utilize usage guarantee or receive discounts. While membership in the network or an administrative fee may be required, the administering organization sometimes offers services such as telephone support, equipment discounts, and training sessions to supplement their vendor contract options.

As has been seen, there are many variables in an online reference service budget which may be manipulated to fit within a given set of financial circumstances. The sample budgets give concrete examples of two different cost approaches to initiating a service. Even more variables are present in the basic structure of an operating budget, as was seen in figure 3.

A thoughtful budget may also include peripheral cost elements which have not yet been mentioned. A selective dissemination of information (SDI) service providing clients with regular search updates could be part of the budget structure. Ready-reference use of online systems may be budgeted for. Document delivery, which may include interlibrary loan, photocopying costs, or online ordering expenditures, is another potential budget element.

Whether or not to include a given cost area may depend on the aspect of budgeting not addressed here: source of income. Whether funds for a given budget item come from a larger institutional budget or from fees charged to clients and patrons may decide the issue. Complete cost recovery through fees of a given service component (search sessions, SDIs, or documents ordered online) will create flow-through money which need not appear in the budget structure. But a meticulous financial management scheme will register all aspects of a service, including sources of income and all expenditures. The extra work involved only contributes to a more thorough, controllable, and flexible online reference service.

Notes

1. Paula M. Strain, "Adopting a Budget: A Primer of Preparation," in Alice Bruemmer, ed., *Library Management in Review*, (New York: SLA, 1981), p. 15.
2. Jean K. Martin, "Preparation of Proposals for Online Bibliographic Services in Academic, Government, and Industrial Libraries," in Ellis Mount, ed., *"Planning for Online Search Service in Sci-tech Libraries," Science and Technology Libraries*, vol. 1, no. 1 (Fall 1980): 12.
3. John C. Blair, Jr., "Considerations in Planning for an Information Center Microcomputer System," *Online Terminal/Microcomputer Guide and Directory 1982-83* (Weston, Conn.: Online, 1982), p. 21.

Bibliography

Atherton, Pauline, and Roger W. Christian. *Librarians and Online Services.* White Plains, N.Y.: Knowledge Industry Publications, 1977.

Bruemmer, Alice, ed. *Library Management in Review*, New York: Special Libraries Association, 1981.

Chen, Ching-Chih, and Susanna Schweizer. *Online Bibliographic Searching: A Learning Manual.* New York: Neal-Schuman, 1981.

Drinan, Helen. "Financial Management of Online Services—A How-to Guide." *Online* 3, no. 4 (October 1979): 14–21.

Fenichel, Carol H., and Thomas H. Hogan. *Online Searching: A Primer.* Marlton, N.J.: Learned Information, 1981.

134 Management

Hoover, Ryan E., ed. *The Library and Information Manager's Guide to Online Services*. White Plains, N.Y.: Knowledge Industry Publications, 1981.

Knapp, Sara D., and James Schmidt. "Budgeting to Provide a Computer-Based Reference Service: A Case Study." *Journal of Academic Librarianship* 5, no. 1 (March 1979): 9–13.

Lee, Joanne H., and Arthur H. Miller, Jr. "Introducing Online Data Base Searching in the Small Academic Library: A Model for Service Without Charge to Undergraduates." *Journal of Academic Librarianship* 7, no. 1 (March 1981): 14–22.

Matzek, Dick, and Scott Smith. "Online Searching in the Small College Library—The Economics and the Results." *Online* 6, no. 2 (March 1982): 21–29.

McClure, Charles R. "A Planning Primer for Online Reference Service in a Public Library." *Online* 4, no. 2 (April 1980): 57–65.

Mount, Ellis, ed. "Planning for Online Search Service in Sci-tech Libraries." *Science and Technology Libraries* 1, no. 1 (Fall 1980).

Online Terminal/Microcomputer Guide and Directory 1982–83. Weston, Conn.: Online, Inc., 1982.

Saffady, William. "The Economics of Online Bibliographic Searching: Costs and Cost Justifications." *Library Technology Reports* 15 (September–October 1979): 588.

Shirley, Sherrilynne. "A Survey of Computer Search Service Costs in the Academic Health Sciences Library." *Bulletin of the Medical Library Association* 66, no. 4 (October 1978): 390–96.

Watson, Peter. "The Dilemma of Fees for Service: Issues and Action for Librarians." In *ALA Yearbook* xv–xxii. Chicago: American Library Association, 1978.

JOHN EDWARD EVANS

Methods of Funding

The advent of online information retrieval services in libraries promises improved and expanded reference service options unlike any other single advance in recent experience. Whereas the large number of printed publications and the desire for greater access to information first prompted the automation of information retrieval activities, now it is too often fiscal inadequacy that determines the availability of such services. The availability of funding options and levels of finance most necessarily take a central role in the operation, indeed, the very existence of the services itself.

Donald King has provided a sharp focus on the problems of financing library services by citing the case of the academic library which has been experiencing restricted budgets and rapidly increasing costs.[1] These problems, combined with a decreasing user population and a disturbing trend in smaller rates of budget increase within the university itself, has made the library in academia especially vulnerable to budgetary restrictions and service development delays. Similar problems are confronting all public, budget-based libraries.

Drake and Olsen confirm that funding alternatives may be bleak as "Libraries must compete with other programs for funding at all levels . . . [and] the amount of money that policy-makers are willing to commit to libraries may be severely limited."[2] Inadequate funding is a characteristic of the times in which we work. The options suggested in this paper are in no way intended to be an exhaustive listing of all the particular methods potentially available for funding an online search service. This work is designed to introduce and examine certain salient funding concerns and demonstrate some successful means of meeting these needs. Locally unique problems may find resolutions in equally unique solutions. Perhaps the most consistently successful group of fund raisers in the library environment are those who can creatively step outside the rigidity of the planning-justification-system-budget-installation scheme for service development.

Before considering the varieties of funding options and methods, it is important to sketch an overview of the costs of online search service operations. In the previous chapter, Grimes provided a thorough cost analysis which is more than adequate to illustrate the types of costs involved in initiating these services. Many libraries can forego certain purchases and options for service either initially or later on as the service demand and user response more clearly define the needs in this area. For a library to begin offering online information services the minimum investment would probably be approximately $2000.[3] Libraries can and do spend far in excess of this amount. Some libraries need staff, some need equipment, some need searcher training, and so on. A moderate, reasonably thorough first-year service will require funding to a level of roughly $5000.[4] If we consider that an online service can be started for $5000 dollars or less, and if that figure is compared with general library expenditures for supplies and equipment, we see that these costs for a terminal and related equipment are relatively small and, in some large libraries, hardly noticeable.

For an online service there are essentially two principle monetary concerns: (1) the initial, or start-up, costs, and (2) the means by which the service is financially maintained on a continuing basis once it is operational. These two aspects will be discussed separately because they reflect entirely different types of problems.

Initial Funding for an Online Service

Only those direct, out-of-budget costs related to online searching will be considered here. This decision is in no way meant to imply that staffing costs are unimportant or minimal; on the contrary, the value of staff-contributed labor probably would far exceed the actual dollar amount needed for equipment and supplies. However, the expenses for staff time, whether in planning or operational phases, need not be considered as integral to the initial funding and budgetary procedures. Hitchingham has questioned the inclusion of staff time and cost recovery even during operational phases[5] and I would extend this to the planning phase as well. Planning for service development as well as training, research, and management activities are part of the traditional activities of librarians.

Implicit in the analysis is the notion that staff will be diverted from traditional work to assume the responsibilities, in whole or part, of online services. This is the most common method, with reference librarians and others being retrained to perform online service roles. In the event that a reference department is in the online service area, the organizational budget would show greater impact in view of salaries, benefits, and services.

What seems most problematic about the initial funding of an online service is characterized best by the operating circumstances of the budget-based institution. Briefly reviewed, the budget-based institution must work within fixed budget figures established in advance by planning and review procedures with regard to the goals and objectives of the organization. Such is the case with most public institutions. Budgeted items are generally grouped in line items for equipment, supplies, and salaries. The line-item allotments are constructed from each activity's plans for the fiscal time period involved. Such budget procedures place some important and valuable constraints on service operations but can be quite problematic when new ideas, technologies, or service options are proposed. This type of budgeting, especially in times of financial retrenchment, admit to little, if any, variation from the budgeting plan of the former fiscal period.

While the costs for online services themselves may not be great in terms of salaries, resource budgets, or general equipment purchases in any given budget year, it is the problem of the unplanned expenditure that causes most fiscal officers to hesitate. The proposal to provide and maintain online services is too often made in the midst of a budgetary cycle, without benefit of previous planning. When officers discover that the costs are too high to be accommodated within the existing budget, they conclude that the service is too expensive. This practice and its concommittant result are attributable to bad planning and procedures. Clearly the solution is to plan for the first-time expenditures adequately and include them in budgetary requests.

Library budgets and the purchase of data terminals for online access speak directly to the issue of initial funding. Terminals, along with searcher training and support documentation such as search aids, represent the major costs in the initial funding needs, with the terminal purchase being probably the single greatest necessary purchase. These expenditures are most generally covered out of the library's budget or through special, one-time funding.[6] These new purchases are not annually repetitive, so it is not necessary or even desirable to seek continuous line-item budgeting. The useful life of a terminal can extend to three years or more; its replacement can and should be anticipated and planned for with future equipment purchases, but other supplies and materials such as search aids and printer paper, purchased on an intermittent basis, can be met from other budgeted categories. With a separate online service budget there are some advantages, such as cost-control, cost recovery through patron charges, and no additional strains on existing budget divisions. But as suggested here, these costs can be covered in existing budget categories.

The cost of the terminal is likely to be the greatest single expense that needs to be met out of budget. There is a considerable variety of terminal models from which the selection can be made. Numerous case studies and guides for terminal selection are available. Implicit in this literature is the suggestion that these costs, substantial to a point, are manageable out of existing funding sources, namely the library's budget, if the purchase has been planned for. Even under extreme budget restraints, a data-terminal purchase could easily substitute for the purchase of another electronic typewriter. Similarly, support documentation and other search aids could be purchased from book budgets, and staff training costs could be brought under professional development, travel, or continuing-education budget categories.

In certain instances, the start-up costs are not available within the institution's own budget and are not forthcoming in foreseeable budgets. Though increasingly unlikely because of declining economic conditions in general and governmental policies and practices, outside funding has been used as a source for these initial expenses. Lake Forest College was a recipient of grants from the Council on Library Resources for a program of database use for college undergraduates. These grants provided start-up cost support and provided for some continuing operations.[7] As an additional example, in 1977 the South Dakota State Library purchased and installed data terminals and related telecommunications devices in the three large university libraries in the state as well as the state library.[8] The total expenditure was less than $5000 for each university library. This level of program support is remarkable for its implications. The university libraries involved had previously relied on remote searching by search analysts at the Bibliographical Center for Research in Denver. This service, too, had been supported by the South Dakota State Library. As demand grew, it became more feasible simply to provide the equipment to the libraries and to let them do their own work and handle their own costs after initial setup.

A second benefit of this plan was that the principal resource libraries in the state were technologically linked for improved interlibrary communication for interlibrary loan and message switching. Such coordinated use of equipment and the resulting benefits provide examples of the extended uses of expensive equipment. These examples are not lost on cost-conscious library administrators. Furthermore, in the South Dakota example, this installation and use of the data terminals allowed the removal of the slow, laborious TWX terminals previously used for interlibrary communication. In this case the installation of expensive database access equipment actually saved money by eliminating the TWX lease.

Other instances of first-time funding outside the library have included special budget requests handled either within the institution and drawing on the organization's discretionary funds or as part of the institution's special request to the funding agency. Some libraries have successfully included special one-time funding requests in their annual budgets. This is an appropriate and not uncommon type of request which could be used to provide funds for the bulk of initial costs. Libraries large and small use this practice to provide OCLC installations, circulation system enhancements, theft detection devices, and audiovisual equipment. Such seed-money allocations are quite applicable to online services. The intent is to provide one-time, start-up funding after which the organization is expected to follow-through with its own budgetary adjustments to provide continuing support. These funding sources may be found within the library's own budget or as part of the larger organization's special expenditures. University budgets characteristically have discretionary accounts and smaller municipalities supporting public libraries use such one-time funding methods to provide initially for new or extended services.

Levin provides an excellent review of the funding process in the case of a new library/information center where initial costs were covered for some projects and other costs met by specific allocations for use. This article also correctly describes the need for planning and development and the pitfalls of working within the budget cycle.[9]

Another example of special funding sources is college endowment income. These funds are sometimes considered discretionary in nature and require only the request, the commitment, and the decision to use them for a specific purpose. One highly unusual instance of outside funding involved an academic library administration which had repeatedly failed to provide funding for online services. The faculty of the university successfully petitioned the university administration for funding for the project. Endowment income was made available to the library administration with the directive that the money was to be used for establishing online services. The total cost for complete system development was $5000.

Such windfalls may not be common but are very useful and appropriate for service introductions such as online reference service. Various friends-of-the-library organizations should be considered as potential sources for these unique gifts. Private philanthropy has been solicited by libraries for years to provide new furnishings, buildings, and special rare-book purchases as well as equipment; the mutual benefits of tax deductions for the giver, and the boon for the library, are obvious. It may not be possible to accomplish a total service development with such income, and the library may need to contribute a part of the necessary funding in a matching-funds arrangement or some similar shared contribution. Sharing costs in this way could be provided out of other budget categories as mentioned above: book funds, travel, or equipment purchases.

Some libraries can make use of existing equipment. Virtually all dial-access, remote data terminals can be used for online searching. There are limitations, but many machines can be adapted for this purpose. Libraries already using dial-access terminals for OCLC cataloging can use the same equipment for database searching. In certain environments characterized by low demand for cataloging or information retrieval or both, such cooperative equipment sharing could prove very cost effective.

Costs can be distributed among several public and technical service operations. A further benefit of sharing costs is that superior equipment can be used at a reduced departmental cost. In so doing, maximal use of expensive equipment is possible, and introducing service enhancements as part of cooperative ventures reduces initial program-specific costs and can pay dividends later when equipment upgrades are necessary. Library administrators can be persuaded by actual activities more forcefully than by probable future demands, no matter how well documented elsewhere.

Sharing equipment can cause problems, most of which concern terminal access time for different applications. Cooperative staff/application scheduling is essential for successful multi-use. Search analysts, catalogers, bookkeepers, and interlibrary-loan staff must understand the others' needs. Flexible scheduling and staff planning can be effective solutions to potential conflicts.

Another source of funding available to some libraries is through the cancellation of some of the higher-priced printed indexing and abstracting services. Such moves are radical and drastic funding alternatives, with certain drawbacks which need consideration. First, in regard to services, there exists the reluctance to curtail a resource and service option known and presumably used by library patrons, in order to provide funds for special projects which for some are unproven and certainly unfamiliar. Further, before suggesting cancellation of subscriptions or standing orders, it should be determined that these funds will be used for the online service. Some libraries operate under strict rules that prohibit the transfer of funds among the categories of salaries, resources, and supplies. Because subscriptions to printed indexes are annual recurring costs, as funding sources for online searching they are probably best applied to the continuing costs of operational expenses. As initial funding they may defray some of the first year's training and resource-gathering expenses. Responses to this question of applying cancelled subscription monies to online searching are favorable and suggest that this means of support is quite workable.[10] Second, in regard to oversight panels or budget officers from whom fiscal approval is required, the library risks implying that the service to be cancelled was unnecessary and general cutbacks can be forthcoming in other areas or overall. Cancelling subscriptions can be the source of not only financial but philosophical issues as well. The long-term results of cancelling subscriptions in favor of online access to the database can aggravate the fee-or-free service dispute anew.

Producers of some abstracting and indexing services who also produce comparable databases are not unaware of these temptations and practices. Consequently, some impose a multitiered price structure based on continuing subscriptions to the printed indexes. Libraries that retain printed indexing or abstracting services may receive discounts for online search royalties, or libraries that terminate print services may receive online surcharges. The differences in these online charges can be substantial. Although recent comments suggest cancellations as a viable alternative for individual libraries, the widespread occurrence of such actions could provide serious problems for libraries and patrons alike. Gardner and Wax have correctly written that such actions could "jeopardize the financial base of the database producers" resulting in subscription requirements or substantially increased royalty fees.[11] The indexing/abstracting industry is not yet sufficiently directed toward online databases to allow stability of costs in this regard. Cancellations alone will be insufficient to subsidize all searching or even all of the first-time costs, even though savings of $1500–$3000 have been reported.

An interesting consideration that parallels the problem of funding through cancellation of printed materials is that by providing online services, a smaller-sized library, which could not have afforded certain print sources, can now provide access to specialized databases. Matzek and Smith describe the circumstances whereby their library, through its online searching, can provide current literature surveillance to science students who previously had little or no bibliographic access through the library because of past fiscal reductions or increased costs.[12] This point specifically supports the position that online searching does not eliminate library service options, rather it increases the number and scope of services available and in some cases provides that which was previously unavailable. The effect of introducing online services in their library is reported by Matzek and Smith as doubling their bibliographic access.

Operational Funding

After the library has an online information retrieval service, the focus of the funding problem shifts to finding the means of providing the necessary financial support and fiscal management to assure continued operation. Operational funding needs center on the costs incurred in the actual conduct of online searches. For the purposes of this paper only those direct costs for online time will be considered, such as computer connect time, telecommunications charges, database royalty charges, and charges for online of offline printing of citations. A survey of the literature on online service costs reveals several articles that enumerate the full costs of an online service that an organization can expect both directly and indirectly.[13]

Indirect costs include staff time or work-hours lost because of other duties, overhead equipment depreciation, increased demand on other services such as interlibrary loan, ongoing staff training, and costs of other related activities such as bookkeeping, fee collection, publicity, management, and future planning. Though far from insignificant, these costs can be seen as (1) part of existing structures and procedures, (2) balanced by staff time, reductions afforded by increased online searching, or (3) of greater significance after the service is flourishing and not of serious consequence in the early phases of initial service operations. This paper does not attempt to examine these specific concerns in detail. For some libraries these costs will be immediately significant, for others they may never be a concern.

The library must consider, however, the cost of each search; specifically, the charges that will be received in the form of monthly invoices from the database vendor/supplier. As part of the initial planning for the online service the library must decide how to provide for the search-specific costs of online searching. The library will decide either (1) to subsidize all costs or (2) to make some attempt at cost recovery, up to and including all costs. To subsidize searches to their fullest, the library will need to be sufficiently well financed to absorb all costs from the the search service. Costs to be encountered here will range from low to high, depending on the type and size of the library, the number of searches performed, the skill of the search analysts, and the costs of the specific databases likely to be used. Actual costs reported in the literature vary considerably. Saffady projects budgets ranging from $8397 to

$41,590 reflecting both ends of the anticipated cost spectrum and covering more than direct costs.[14] From Raedeke's one-month test of public library operations an estimated $8800 annual search cost is inferred.[15] Cogswell anticipated annual expenses of at least $10,000 for searching alone, in an academic library.[16] Knapp, with considerable data for support, projected annual costs to the library in the $15,000–20,000 range for direct expenditures.[17] Firschein et al. estimated a public library service to cost approximately $14,000 annually.[18] Drinan, offering an excellent example of fiscal assessment, foresaw corporate library expenses to slightly exceed $28,000 each year.[19] Crawford and Thompson do not offer dollar amounts but indicated that their academic library would absorb the costs because "not charging for online searching is a feasible alternative, and a necessary one if the service is to be treated effectively as an integral part of reference services."[20]

For those libraries which cannot provide a full subsidy of $5000–$12,000 a year, cost recovery in whole or part is the most viable alternative. Although there seems to be considerable variety to the cost-recovery options available, they tend to (1) full subsidy with no charge to the patron, (2) partial subsidy with the balance paid by the patron, and (3) cost recovery from the patron with little or no direct or indirect subsidy. While each mechanism is workable, each is also accompanied with certain problems, some of which are detailed below. It is important at this point to underscore the need for precise, well-developed policies for operational and fiscal management and the appropriate procedures to support these policies. Drinan's work provides an excellent example of the rationale and methodology of budgeting and forecasting.[21] This work, the extensive modeling by Casper[22] and the examples provided by Saffady[23] would be helpful for those preparing to introduce an online service. What is paramount is that the practices, methodologies, and models used should be so constructed as to correctly reflect the goals and objectives that originally prompted the system planning.

Full-subsidy searching is the cleanest policy and the easiest to administer; one simply does the searches and distributes the results. It may also be the hardest economic option with which to live. Applying few, if any, criteria to the decision to do a search, and not asking patrons to apply realistic limits to their information needs can result in an overwhelming demand for service. This demand can result in a collapse of the service, at worst, or simply cause time and paperwork overloads resulting in unacceptable delays. With no financial restraints on patrons or search analysts there needs to be an adequate mechanism by which searchers are limited in number, scope, or by some measure of applicability. There may very well be no limit to the amount of time or money that can be expended on searches. Even the wealthiest library will eventually need to set some limits on the activity. A policy of full-subsidy searching requires that these limits be defined in anticipation of the service to ensure that, once operational, the service will be continuing and effective. Such a policy, though easily developed and clearly stated, can cause enormous problems for the library. It is not uncommon for a library to spend many thousands of dollars in support of this service, and such charges are nearly impossible to budget in advance.

There is usually an increasing demand for other library services, such as interlibrary loan, following the installation of online services.[24] Changes of this sort will impose increasing fiscal and service pressures that few libraries today can easily accommodate. It is not surprising, then, that libraries that adopt this form of funding usually limit service based on various aspects of the following: patron status, quotas, search applicability, printed sources available, tolerated time delays (intentional or unintentional), search or printout length, and others.

Partial-subsidy plans attempt to bridge the policy gap perceived by many between the notions of free and for-fee information services. Often those libraries which employ such a scheme are asking the patrons to assume at least partial financial responsibility for their unique and specialized information-gathering activities.[25] Partial-subsidy policies can be easily stated and introduced to the user community accustomed to free services. Although any cost-recovery activity tends to create another level of bureaucracy within the library to handle the receipts, a partial-subsidy, or token-payment plan for users, represents one of the simplest bookkeeping schemes. Libraries find this method at least partially limits user demand. Budget projections are not likely to be accurate, but expenses are greatly reduced when compared with full subsidies. Nevertheless the library's expenses can be high if not regulated and are still relatively unanticipated.

Some of the more popular variations on partial subsidy include: (1) dividing the costs of the search according to a fixed percentage, for example, 50 percent paid by the patron and library each; (2) a basic fee of $1, $5, or $10 paid by the individual, the balance paid by the library; (3) dividing the cost evenly to a certain dollar amount with the patron paying the full cost beyond that predetermined limit; (4) a credit for each faculty member or department with additional costs paid by the individual; and (5) a basic fixed-dollar subsidy for each search, the balance paid by the patron. Under these schemes the patron, the library, or both stand to incur considerable costs. The budget and bookkeeping in these cases can be awkward.

A further disadvantage of subsidy and partial-subsidy plans is that gross, unqualified searching practices may develop which fail to provide or elicit the relevant literature to respond to the information need. This disadvantage is less obvious than the impact of search costs on the library but may be more disastrous in the long run. When costs are so openly subordinated to the considerations for the service, the search analysts are relieved of pressures for cost-effective performance, which is to say that the searcher's accountability is obviated. For those who supervise the performance of information retrieval services and the responsible staff, this sort of hand washing is as unacceptable since it punishes errant searchers for failed searches by having them pay for their mistakes. There exists the assumption that strategy and performance failures can be absorbed. When such problems arise, that is, high costs, low productivity, and less-than-valuable or even unusable results, user confidence in the service can drop. When such a deterioration of service reliability occurs, regaining the patron's confidence may be difficult at any price. In one case with which I am familiar, this sort of policy resulted in a 38 percent drop in service demand in the first year, despite a generous subsidy plan.

The remaining cost-recovery variation is that of charging users for the full cost incurred in the individual search. Perhaps as clear a policy statement can be written here as with the full subsidy. Each patron is charged, in effect, by individual interests and demands. Search analysts are under a certain degree of pressure to provide efficient and cost-effective searching. This practice provides a self-regulating demand based on the user's needs and desires. The library's costs are regulated to a degree, in that the costs of search activity are directly recovered. The library, though, will still incur some expenses, beyond the basic online charges, in the form of unrecoverable costs for administrative use of the service. Some libraries compensate for this by imposing minimum charges or by adding surcharges keyed to actual activity; for example, each patron may be charged a fixed percentage beyond the direct cost. This policy has been successfully employed for two years at Memphis State University libraries. Overall the service has posted significant gains in user demand (187% and 28% net increases in the two years respectively).

Cost-recovery plans immediately evoke discussions of user fees. The argument over user fees in the literature reflects in large part the problems of funding which are to be encountered throughout the contemporary library scene, that is, if traditional funding sources were still available or sufficient we would not now be concerned with user fees. User fees are an interesting problem and a significant barometer of current information center funding levels.

The imposition of fees for online searching is for all too many libraries an absolute necessity. If the library did not recover the costs of searching, the searching could not be done. Or, in the case of partial subsidies, if the library did not recover the balance, the service would not be available. Libraries have for many years been forced to charge for interlibrary loan and photocopy services. As I have not yet enjoyed the luxury of any of these services for free, I judge charging user fees lamentable, but unavoidable all the same.

It should be understood that all of the aforementioned options for operational funding are workable and have been found acceptable by library users. The critical factors in each case are what the library can afford and chooses to do. The realities of the situation should work in concert with the policies. Failing to pay attention to the realities can result in dismal failures and there are entirely too many cases where this has occurred.

All fiscal management schemes will create an accompanying bureaucratic hierarchy to insure that the service continues to be viable. Cost-recovery schemes evoke several layers of such bureaucracy. The department head or manager must have adequate information on expenses and income to maintain budgetary accountability. This information is provided by regular reports on service activity from the principal online search specialist. Log sheets detailing precisely and thoroughly all search activity are essential not only for cost recovery but for evaluation as well.

The handling of search requests, the distribution of the results of the searches and the receipt of payments are commonly centralized. Fee collection, whether actual payment or instructions on payment methods, is best handled at the time the search results are provided to the patron. In larger organizations like universities and corporations, charges are commonly made to departmental, grant, or special-fund accounts. This method of payment is usually handled by paperwork transfers among accounts. In those instances when individuals are paying, personal checks or other forms of payment are accepted. These activities are commonly performed by clerical staff under the supervision of professional librarians. In many libraries the overall bookkeeping functions, for online searching including cash deposits and bill paying, are subsumed by the library's own staff as simply another account. These activities are by no means difficult nor should they be viewed as problematic. Systems and procedures will in large part suggest themselves in each library, and similar mechanisms may already exist.

Throughout the development of each library's plans for and conduct of an online information retrieval service constant attention must be paid to several factors. There must be a thorough understanding of the goals and objectives for the service. These goals and objectives must be realistic and take into consideration the budget, cost projections, user demand, policy implications and, above all else, the service interests of the library and the patrons to be served.

Notes

1. Donald W. King, "Pricing Policies in Academic Libraries," *Library Trends* 28, no. 1 (Summer 1979): 47-62.
2. Miriam A. Drake and Harold A. Olsen, "The Economics of Library Innovation," *Library Trends* 28, no. 1 (Summer 1979): 96.
3. Carol H. Fenichel and Thomas H. Hogan, *Online Searching: A Primer* (Marlton, N.J.: Learned Information, 1981), p. 79.
4. Charles R. McClure, "A Planning Primer for Online Reference Service in a Public Library," *Online* 4, no. 2 (April 1980): 57-65.
5. Eileen E. Hitchingham, "Use of Commercially Vended On-Line Data Bases by Academic Libraries," in *Proceedings of the American Society for Information Science* 12 (1975): 159.
6. David M. Wax, "A Handbook for the Introduction of On-Line Bibliographic Search Services into Academic Libraries," Office of University Library Management Studies Occasional Paper number 4 (Washington, D.C.: June 1976), pp. 28-29.
7. Joann H. Lee and Arthur H. Miller, Jr., "Introducing Online Data Base Searching in the Small Academic Library: A Model for Service without Charge to Undergraduates," *Journal of Academic Librarianship* 7 (March 1981): 14-22.
8. Herschel V. Anderson, "Online 'free': How they do it," *Library Journal* 104, no. 14 (August 1979): 1498.
9. Ellen J. Levin, "Establishing a Special Library the First Year," *Special Libraries* 73, no. 3 (July 1982): 193-201.
10. "Action Exchange" *American Libraries* 13, no. 10 (November 1982): 619, 621.
11. Jeffrey J. Gardner and David M. Wax, "Online Bibliographic Services," in *Reference and Information Services*, Bill Katz and Andrea Tarr, eds. (Metuchen, N.J.: Scarecrow, 1978), pp. 244-5.
12. Dick Matzek and Scott Smith, "Online Searching in the Small College Library—The Economics and the Results," *Online* 6, no. 2 (March 1982): 21-29.
13. William Saffady, "The Economics of Online Bibliographic Searching: Costs and Cost Justifications," *Library Technology Reports* 15, no. 5 (September-October 1979): 567-636; Helen Drinan, "Financial Management of Online Services—A How-to Guide" *Online* 3, no. 4 (October 1979): 14-21.
14. Saffady, "Economics of Online," 599-600.
15. Amy Raedeke, "Machine-Assisted Reference Service in a Public Library: A One-Month Test Period," *Online* 2, no. 4 (October 1978): 56-59.
16. James A. Cogswell, "On-Line Search Services: Implications for Libraries and Library Users," *College and Research Libraries* 39, no. 4 (July 1978): 277.
17. Sara D. Knapp and C. James Schmidt, "Budgeting to Provide Computer-Based Reference Services: A Case Study," *Journal of Academic Librarianship* 5, no. 1 (March 1979): 9-13.

18. Oscar Firschein, Roger K. Summit, and Colin K. Mick, "Use of On-Line Bibliographic Search in Public Libraries: a Retrospective Evaluation," *Online Review* 2, no. 1 (1978): 41–55.

19. Drinan, "Financial Management," p. 19.

20. Paula J. Crawford and Judith A. Thompson, "Free Online Searches Are Feasible," *Library Journal* 104, no. 7 (April 1979): 295.

21. Drinan, "Financial Management," pp. 14–21.

22. Cheryl A. Casper, "Pricing Policy for Library Services," *Journal of the American Society for Information Science* 30, no. 5 (September 1979): 304–9.

23. Saffady, "Economics of Online," passim.

24. Anderson, "Online Free," p. 1498.

25. Drake, "Library Innovation," p. 96.

JAMES RETTIG

Options in Training and Continuing Education

Just as one must receive education and training before offering reference service, one needs at least some training before conducting online searches. Given this absolute, how, where, and from whom does one receive the necessary training? A number of options exist which will be discussed in this paper (see table 1). Most of them, such as library-school courses, workshops, conferences, books, journals, and vendors, are already familiar to librarians as sources of education for other activities. Each has advantages and disadvantages; the experience of many librarians has proved that these various kinds of education should be combined with each other.[1]

Training to simply search is not enough if a library's online reference service is to be a success. Education on the uses of online services must come before training to search. All the staff require education so that there will not be "an isolated, elite cadre of searchers whose talents and resources are not utilized by other librarians."[2] All the staff must understand what new service options online capability creates and how they can exploit databases in their work in administration, acquisitions, cataloging, serials, and government documents. Those reference librarians who are going to take search requests from clients, even if others will actually do the searches, require additional education about the intensive reference interview that precedes the search.[3]

The library's (ideally the parent institution's) administration requires education on the possibilities and the limitations of online systems. (In one situation, an academic library director, upon being disabused of the notion that DIALOG was a full-text online delivery system that would permit a drastic reduction in serials subscriptions, cancelled plans to offer service through DIALOG.) In addition to the advantages and limitations of online service, administrators "should know the basics of online searching in order to understand how much training is necessary, what kind of equipment ought to be used, and the stress under which

TABLE 1. Sources of Education

METHODS	ADVANTAGES	DISADVANTAGES
SELF-INSTRUCTION	Relatively inexpensive	No interaction with others Very easy to form mistaken ideas and base searching behavior on them All online practice at library's expense
LIBRARY SCHOOL COURSES	Thorough presentation of theory Trainees have ample opportunity for interaction	Too extended for practitioners Availability limited by location Online practice time sometimes limited
INDEPENDENT TRAINERS	Short time span Good online environment Interaction with other trainees	Combined cost of travel and registration Instructor not available afterwards Lack of online practice time afterwards
NETWORK TRAINERS	Short time span Cost comparable to vendors' Establishes relationship with network personnel Training sites close to trainees Trainees observe each other	Instructor not available afterwards Lack of online practice time afterwards Combined cost of travel and registration
VENDORS' TRAINING	Short time span Variety of sites to choose from Trainees observe each other Includes online practice time afterwards	Instructor not available afterwards Some important topics often ignored
IN-HOUSE TRAINING	Instructor on-site afterwards Can be individualized No travel or registration costs	Costly and time-consuming to prepare All online practice at library's expense Possibility of personality clashes

NOTE: These major advantages and disadvantages *generally* apply to each method; there may be exceptions in particular programs.

each searcher works."[4] This cannot be overemphasized. Training is every bit as important as the service itself since the quality of the service delivered is largely a product of the searcher's ability. In fact, the absence of good training has been called "the largest obstacle in full utilization of on-line information retrieval systems."[5] Training inevitably involves a certain expense and a library's administration must commit itself to supporting training at the same time it commits itself to providing the service. Anyone responsible for providing general education to staff colleagues can readily be educated through literature about online services and by talking with vendors and librarians from other libraries.

Training librarians to conduct searches is more difficult since so many options exist. Through reading and trial and error one can, of course, attempt to teach oneself database selection, search strategy design, log-on procedures, system commands, and logical operator functions. In the early years of online systems this was, of necessity, a fairly common training method.[6] This method has the distinct disadvantage of isolating the would-be searcher from others experiencing the same problems. One searches in vain for any praise of this method.

Library Schools and Intensive Workshops

Library schools in recent years have offered separate courses in on-line reference work or incorporated it into other courses. Even with vendor discounts, the costs of offering adequate lab time has posed a problem for the schools since lab time has been judged, "on a per student basis," as "probably the most expensive lab activity on campus (other than the use of linear accelerators of large telescopes)."[7] Few practicing librarians, save those working close to library schools, can take the time for such a course whether it be for several weeks or for a fourteen-week term.[8] These courses have one undeniable strength. They can cover more thoroughly than other methods the theoretical background of machine-readable file structure and its implications for searching. These courses also tend to differ from vendor and database-producer programs, which concentrate "on helping their customer to know enough about the conventions of their system to be able to use their products properly": the courses are more often "concerned with the on-line experience in the service of providing students with a body of intellectual skills and conceptual frameworks . . . to deal with systems of the future as well as those today."[9]

Independent workshops such as those offered by the University of Pittsburgh's On-Line Training Center compress training time from weeks to days.[10] These workshops immerse students in the online environment. They are expensive, however; travel, meals, and lodging plus registration must be paid. Some networks which serve as brokers between vendors and libraries also offer introductory training. Their trainers travel through the network area, offering training at a cost lower than or equal to that charged by vendors.

Vendor Introductory Seminars

Vendor introductory seminars are a very common form of initial training. A day or two in length, the seminars are conducted through the year in numerous locations, usually costing less in travel expenses and often less in registration fees than the more concentrated workshops. Vendor seminars combine lectures, printed notes, transparencies, and terminal practice on the full system. They concentrate on system-specific features and commands rather than on the databases for particular subject areas. Students generally observe others on the terminal during practice time. This practice enables them to consult each other and to learn from each other's mistakes, a sort of learning that has strong advocates.[11] The instructor is also present to answer questions and offer advice. The instructors' knowledge of their system is, of course, up-to-date.

Participants in vendor seminars usually receive free time on the system, to be used within a specific period after the training session. All agree that the best way to gain knowledge of a system and proficiency at using it is to use it, the more the better. When free time is available after training, administrators should make sure that the eligible personnel use all of it. This may require temporary changes in work schedules or assignment of duties.

Critics of vendor training point out that such seminars sometimes verge on a sales pitch and that important items which are not system-specific, such as how to use the telecommunications networks to reach the vendor's computer and specific database characteristics, are often skimmed over or ignored.[12] Training ten to fifteen people makes individualized instruction difficult and after the training session the instructor is not present for consultation. However, all three major vendors (DIALOG, SDC, BRS) and many database producers offer assistance through toll-free telephone lines.

If no one on a library's staff has experience in any of the major systems, at least one librarian—one who will conduct searches and who possesses teaching ability—should attend an appropriate program offered by a library school, independent trainer, network trainer, or vendor. If a new system is being added, an experienced searcher may be able to learn it through self-instruction; however, the conditions under which self-instruction can succeed are not yet certain.[13]

In-house Training

When an experienced searcher is already on the staff, in-house training of others may seem like the most economical method. Conducting a successful in-house training program is like a tightrope act—everything must be in balance. If only one or a very small number of the staff are to receive training, the interaction of their personalities with the trainer's must be considered.[14] Just as not every librarian makes a good searcher, not every searcher makes a good trainer. If prospects are good, the program can begin, but not until the trainer has selected or developed teaching materials and a training schedule has been set up.

Successful in-house programs exhibit a high degree of structure and discipline.[15] Sessions of predetermined length are held at scheduled times. Between sessions trainees must perform specific online exercises, which the trainer checks. Trainer and trainees meet regularly to review progress and discuss problems. The library must, of course, absorb the full cost of online practice since online practice is indispensable for any introductory training. If several people are being trained at once, their constructive criticism of each other's efforts and sharing of mistakes greatly enhances the experience's educational value. In-house training has several other pluses: the trainer is on site even after the initial sessions, some out-of-pocket expenses can be reduced, and, best of all, instruction can be tailored to meet individual needs.

In-house training is good for people who need only a limited knowledge of one or two databases on a single system, such as an interlibrary loan paraprofessional who needs to verify citations. In-house training also seems well suited to large libraries and organizations where there is ongoing need to give new people introductory training. These needs can justify and amortize the considerable hidden costs for time "spent on developing illustrative search strategies and portions of search strategies,"[16] as well as personnel costs for actual training time plus costs of materials. In other cases, just "the computer connect time and the telecommunications charges for in-house practice . . . may . . . exceed the cost of sending an individual to training or of hosting on-site training at a group rate."[17] In-house training demands a trainer with genuine teaching ability whose knowledge of systems and databases approaches or equals that of vendor and database producer training personnel.

Continuing Education

Systems and databases change and grow continuously. While it has not been proved conclusively, none question the need for education and training beyond a system introduction.[18] This, too, is available from a variety of sources.

Several of these sources are very familiar to librarians and include conferences, books, and journals. Conferences or professional organizations such as the American Library Association, the Special Libraries Association, and the American Society for Information Science invariably give some attention to online searching. Others, such as the annual National Online Meeting in New York in the spring and the annual Online Conference in the fall, dedicate themselves to online subjects.

Similarly, journals such as *RQ, Reference Services Review*, and the *Journal of the American Society for Information Science* carry relevant articles; and *Online, Database,* and *Online Review* concentrate on this area. All are important sources of continuing education for searchers. The three major vendors and many of the database producers publish newsletters for their users. These are *must* reading for searchers who wish to keep their knowledge current. With their indexes, the newsletters become reference tools almost as useful as the vendors' and database producers' manuals.

Vendors also offer other forms of continuing education: seminars for advanced searchers on system-specific capabilities, annual updates, and subject seminars on specific groups of related databases. These rarely include terminal practice, but they are worthwhile nonetheless since they impart a great deal of practical information and review system fundamentals. The annual update sessions are the least valuable of the vendors' educational programs since their newsletters publish much of the information offered.

Vendors sometimes offer free online time, especially on new databases. The manager of a library's online reference service should watch for these opportunities and make sure that searchers with appropriate subject expertise get time to take advantage of this practice time.

Database producers often conduct programs in various locations to educate users about their products, sometimes at no charge to participants. These frequently include free online practice during the programs. Like the vendors' subject seminars, these programs offer a close look at the contents and structure of databases. Without such knowledge searchers cannot take full advantage of system capabilities for searching specific data fields or for manipulating the form of output. If a library uses a database or set of databases a great deal, such continuing education is an investment in quality service. Many libraries base in-house continuing education on information gathered by one staff member who attends a workshop and reports back to other library colleagues.[19]

Many online-user groups also provide continuing education. A user group is a group of searchers, either from the same geographical area or who use the same online system, who meet to discuss common concerns.[20] User groups vary considerably in structure but all have the same fundamental purpose—to bring searchers together to learn from one another. Most user groups are loosely structured and meetings tend to follow a similar pattern with members discussing problems, sharing discoveries, asking questions, and swapping information. Much good can come from participating in a user group. Asking a group a question can lead to a more complete answer than one would receive asking just one person, even one expert at the end of a toll-free help line. User groups sometimes sponsor database producer workshops or offer formal programs presented by members. Topics include hardware, database profiles, management issues, and telecommunications. Having to prepare a presentation for one's peers is an excellent way to learn a great deal on one's own.[21] Some groups compile directories that note members' online search aids (e.g., thesauri, database guides, relevant journal holdings), placing help just a phone call away.

The National Online Circuit is a loosely organized umbrella organization for user groups. It has compiled a directory of user groups and hopes to serve as a clearinghouse for their news by using as one of its principal vehicles of communication the *Online Chronicle* (file 170 on DIALOG), an electronic magazine.[22]

In any library where two or more librarians conduct searches, there is, in effect, a small online-user group. Searchers within a library should consult each other freely and exchange tips. Where degrees of expertise or knowledge differ considerably, it might be prudent to institute a buddy system whereby a more experienced searcher checks an apprentice searcher's strategies before the apprentice goes online. Such a system has been used at SUNY-Albany where most of a six-month period a trainee works alongside an experienced searcher, both in interviewing users and in formulating searches. The expert helps the trainee to focus on (1) vocabulary and (2) developing alternative strategies for the same search.[23] Whether or not such an arrangement is a part of a library's formal training or continuing education, searchers may want, on a regular basis, to take a search request and have each member create a list of databases to be searched and appropriate searching strategies. Staff can exchange these, discuss them, and learn of different approaches. One such participant notes that "competition was not a factor and there was a feeling of community. One important reason for the positive atmosphere was the absence of grading or personnel evaluation. Another reason was that everyone took turns so each person was in the same situation."[24]

Although highly structured by system protocols, searching is an individualized process. It is clear that "for many searches there is not one right answer but rather a continuum of possible solutions."[25] Without some public sharing of these different approaches, the typical searcher will think only of the one personally developed approach. Just as supervisors' eyes should be "kept open for the newly-trained searcher who has not been seen near a terminal since completing the first week's exercises,"[26] supervisors should also be on the watch for trained searchers who shy away from continuing education, education which cannot be received strictly by self-teaching any more than can introductory training.

The combination of methods a library selects for training its librarians to conduct searches will depend partially upon its situation. Proximity of training facility (library school, independent trainer, site of frequent vendor database producer programs) will influence the library's choice of training type. Other considerations might be the library's relationship with a broker network or the frequency with which the library will need to train new personnel.

Whatever combination of methods employed, there is a unanimous opinion that instruction programs must always include both printed manuals and terminal practice.[27] In addition to being a part of formal training sessions, practice must also take place afterwards. Interaction with other trainees and with experienced searchers is important and remains important as long as one searches. Until a searcher feels comfortable and demonstrates the ability to work independently, it seems best for this interaction to be structured and include some form of evaluation. Because every trainee brings to the task of learning to search a unique combination of subject knowledge, manual information retrieval ability, and attitudes towards technology, training and continuing education should be as individualized as possible.

Online information retrieval is an important extension of library reference services. To be effective, the people offering the service must be effective. This necessitates introductory training and continuing education for searchers. The online community is interdependent; database producers, database vendors, and database users need each other and should consider each other friends.

Notes

1. James B. Tchobanoff, "The Education and Training of Online Searchers: The Searchers' Experience," in *Education and Training for Information Services in Business and Industry in Developing and Developed Countries* (The Hague: Federation Internationale de Documentation, 1980), pp. 7–29.
2. Greg Byerly, 'Training without Education: A Lost Cause," *RQ* 20 (Spring 1981): 231.
3. Where this practice is common, it calls for special handling; see Mary Hardin and Danuta A. Nitecki, "A New Breed of Online Specialty: The Remote Profiler," *Online* 2 (July 1978): 31–38.
4. Quoted anonymously in Darrell L. Jenkins, "Online Searching and Library Administrators: A Casual Connection?" in Martha E. Williams and Thomas H. Hogan, comps., *Proceedings of the 1982 National Online Meeting* (Medford, N.J.: Learned Information, 1982), 255.
5. Christine L. Borgman, "User Training for On-Line Systems," in Allen Kent, Harold Lancour, and Jay E. Daily, eds., *Encyclopedia of Library and Information Science* 32 (New York: Dekker, 1968–81): 265.
6. Tchobanoff in "The Education and Training of Online Searchers" reports that 65.6 percent of his sample learned at least some of their searching skills through self-teaching yet they ranked this method behind vendor and database producer training programs.
7. Charles P. Bourne, "Selecting and Estimating Equipment & Resources," in *New Techniques in the Teaching of Online Searching. An Institute for Library Educators,* Edmund Mignon, ed. (Arlington, Va.: ERIC Document Reproduction Service, ED 181 935, 1978), p. 105.
8. Sara D. Knapp and Jacquelyn A. Gavryck, "Computer Based Reference Service, a Course Taught by Practitioners," *Online* 2 (April 1978): 65–76, describes an intersession course.
9. Edmund Mignon, "Reflections on Training," in Edmund Mignon, ed., *New Techniques in the Teaching of Online Searching. An Institute for Library Educators* (Arlington, Va.: ERIC Document Reproduction Service, ED 181 935, 1978), p. 94.
10. Patricia J. Klingersmith, *Training Program for Bibliographic Searching* (Arlington, Va.: ERIC Document Reproduction Service, ED 191 453, 1980), describes the creation of the center at Pitt. Rowena W. Swanson, "An Assessment of Online Instruction Methodologies," *Online* 6 (January 1982): 38–53, discusses the programs of several independent training centers. For suggested guidelines for independent workshops; vendor introductory, advanced, and subject training; and database producer training, see "Online Training Sessions: Suggested Guidelines," *RQ* 20 (Summer 1981): 353–75. These guidelines were developed by the Education and Training of Search Analysts Committee of the Machine-Assisted Reference Section, Reference and Adult Services Division, American Library Association.
11. Tchobanoff, "The Education and Training of Online Searchers," pp. 7–8, testifies that in a similar experience, others' "encouragement and constructive criticism was the single most important factor in preparing me to be a competent online searcher."
12. Elaine Caruso, "Training and Retraining of Librarians and Users," in Allen Kent and Thomas J. Galvin, eds., *The On-Line Revolution in Libraries* (New York: Dekker, 1978), pp. 207–28.

158   Management

13. Borgman, "User Training for On-line Systems," p. 300, comments that "It is not yet known to what extent the learning of one on-line search system affects the ability to learn another system with minimal training, or there may be little transference from one system to another. Research may show how best to utilize a searcher's previous experience in training on another system." Yet opinions abound. Joan Leigh and Wesley Simonton, "Education and Training for Online Information Systems," in *Education and Training for Information Services in Business and Industry in Developing and Developed Countries* (The Hague: Federation Internationale de Documentation, 1980), p. 93, state their "belief . . . that the students needed to be entirely comfortable with one system in order to preserve the intellectual component of the search without interference from the mechanics of online activity." Elaine Caruso, "Hands on Online: Bring It Home," *On-line Review* 2 (September 1978): 251–68, believes multiple multidatabase systems should be learned simultaneously. Charles P. Bourne and Jo Robinson, "Education and Training for Computer-Based Reference Services: Review of Training Efforts to Date," *Journal of the American Society for Information Science* 31 (January 1980): 26, say that searchers "need to learn one or more search systems (preferably one at a time). . . . As searchers gain experience, they can learn new databases and search systems rather easily and can make use of sophisticated search techniques." Amidst these contradictions one must agree with Borgman's cautious statement.

14. See Jean E. Crampon, "Training Backup Searchers . . . Or How 'Reluctant Ralph' Deals with 'Terrified Tess,' " *Online* 4 (October 1980): 25–29, for a classification of types of potential trainers, trainees, and their interactions.

15. Programs fitting this description have been described by the following: Nancy J. Gershenfeld, "In-house Online Training of the Information Specialist: A Program to Meet the Needs and Capabilities of the Individual Searcher," in Martha E. Williams and Thomas H. Hogan, comps., *Proceedings of the 1982 National Online Meeting* (Medford, N.J.: Learned Information, 1982), pp. 117–21. John Heyer, "In-house Training for Online Searching at a Special Library," *Online Review* 4 (December 1980): 367–74.

16. Swanson, "Assessment of Online Instruction," p. 42.

17. Kristyn Kuroki, "Training the Searchers," in Ryan E. Hoover, ed., *The Library and Information Manager's Guide to Online Services* (White Plains, N.Y.: Knowledge Industry Publications, 1980), p. 191.

18. Carol H. Fenichel, "Online Searching: Measures That Discriminate among Users with Different Types of Experience," *Journal of the American Society for Information Science* 32 (January 1981): 23–32. Fenichel compared the performance of searchers grouped by length and type of searching experience. She concluded that "the only clearcut differences that could be attributed to experience were that the Novices searched more slowly and made more errors than the experienced subjects," p. 29.

19. Barbara Shupe reports that this is the practice at SUNY–Stony Brook; see David Y. Allen, ed., *The Implementation of Data Base Searching at Three Campuses of the State University of New York* (Arlington, Va.: ERIC Document Reproduction Service, ED 198 834, 1980), p. 22.

20. For an overview of user groups, see Mary C. Berger and Barbara Quint, "Online User Groups," in Ryan E. Hoover, ed., *The Library and Information Manager's Guide to Online Services* (White Plains, N.Y.: Knowledge Industry Publications, 1980), pp. 233–49.

21. Jo Robinson, "Education and Training for Computer-Based Reference Services: A Case Study," *Journal of the American Society for Information Science* 31 (March 1980): 97–104, reports on such an activity and its benefits.

22. For further information about the National Online Circuit, contact Rebecca A. Gonzalez, ITT Research Institute, 10 W. 35th St., Chicago, Illinois 60616, (312) 567-4356, and see her "Circuit News" column in issues of *Online*. See also "User Group Directory," *Online* 6 (January 1982): 82–85.

23. Donna R. Dolan, "Training Search Service Personnel," in *Aspects of Information Service Management. Proceedings of the Fall Meeting of the New England On-Line Users Group* (Arlington, Va.: ERIC Document Reproduction Service, ED 180 453, 1978), p. 43.

24. Robinson, "Education and Training for Computer-Based Reference," p. 103.

25. Ibid. Furthermore, Fenichel, in "Online Searching: Measures That Discriminate," proves that a continuum of possibilities exists for searches. She found "unexpectedly large individual differences in search behavior," p. 30. Judith Wanger, *Evaluation of the Online Search Process* (Arlington, Va.: ERIC Document Reproduction Service, ED 190 143, 1980), after comparing searchers in terms of the relevance and precision of their searches, concluded "that the study results dramatically emphasized for us what has long been hypothesized—the fact that searching is a highly individualized process," p. 6.

26. Gershenfeld, "In-house Online Training of the Information Specialist," p. 120.

27. Swanson, "Assessment of Online Instruction," p. 44. A chorus of assent to this comes from all; a sampling follows: Pauline Atherton and Roger W. Christian, *Libraries and Online Services* (White Plains, N.Y.: Knowledge Industry Publications, 1977), p. 48: "one can only become a proficient online searcher by being online. Learning to search data bases effectively can only be done through individual practice, online, at the computer terminal." Borgman, "User Training for On-Line Systems," p. 295: "No matter which method of instruction is used, printed instructions seem to be a necessary accompaniment." Caruso, "Hands On Online," p. 254: "provision needs to be made for practice time that is sufficient to fix the more mechanical aspects of the search process and to allow for more free exploration to round out the skills learned." Thomas P. Slavens and Marc E. Ruby, "Teaching Library Science Students to Do Bibliographical Searches of Automated Data Bases," *RQ* 18 (Fall 1978): 41: "Many feel that the 'hands-on' experience at the terminal illuminates points which are not clear during the lectures."

MARYJANE S. COCHRANE

Publicizing an Online Search Service

Online searching has become a familiar topic to many librarians, but it is still a new phenomenon to the majority of library users. This makes it a candidate for a good publicity campaign. This paper will examine some techniques for starting and maintaining a computer search service publicity program. Printed materials, demonstrations, nonprint media, and publicity by word-of-mouth will be discussed.

Starting the Campaign

Library users often feel they know all they need to about library services, or they are uneasy about "bothering" a librarian with a question. However, they seem to feel more comfortable approaching a librarian for information about a new or different service such as computer searching. This attitude gives the service a head start with a publicity campaign.

Four basic issues should be considered before the first handout is designed or talk given.

First, the reasons for offering the search service must be defined. What points should the potential users be made aware of—the speed of online searching, its retrieval capabilities, perhaps the contents of a database? The aspects of online searching that best meet potential users' needs and thus, appeal to them the most, must be determined.

Second, target groups for the publicity campaign must be defined. Not every library user is a potential online user. Promotional devices should reflect the different types of users to be attracted. In a university setting, publicity may be mainly directed towards faculty and graduate students, as opposed to undergraduate students. A public library may wish to design some of its promotional materials to appeal specifically to local businesses. The key is to know the characteristics and needs of these different users.

The timing of the publicity campaign must also be considered. Publicity should begin in advance of or coincide with the opening of the service. But be sure of the starting date of the service! One university library widely advertised its search service, only to have its vendor contract delayed. Staff then had to refer requests to another library for several months.

Costs are the final consideration. Publicity does not have to be expensive. However, costs of printed materials, search demonstrations, and other publicity techniques should be calculated in advance. Investigate all in-house sources of assistance which may help keep costs down. Remember, too, that sufficient time must be allowed for planning and producing the publicity items.

Then, after considering these issues, different methods of publicizing the search service can be investigated. Some combination of the aids discussed in the following paragraphs will constitute a good publicity campaign.

Printed Materials

Printed materials such as brochures, posters, and handouts provide a forum for introducing the search service. Brochures should be attractively designed, their message clear and concise. Jargon and technical terms should be avoided or, if necessary, they should be explained. Terms such as *offline, Boolean logic,* and *connect time* don't convey a great deal to many people. A brochure should explain what the service can do for the user, not just list databases or vendor names. A question-and-answer format works very successfully. Design the brochure or handout with the user in mind. If possible, have the brochure professionally printed rather than photocopied or mimeographed. Good ideas can be found by reading the materials prepared by other search services, and they are free.

One-page handouts on a single database or a subject bibliography can be very effective in advertising the service. Handouts should be to the point and attract specialized users. Posters may also be used. Again, the information on the poster should be brief and phrased to appeal to the users. Catch phrases like "Need Help with Research?" or "Subject Bibliographies Made to Order" are intended to let the user know immediately how the search service could aid them. A logo used on all the service's promotional materials is helpful in unifying the material. Giving search service promotional materials a common color and print type also help set them off from other library materials.

Distribution is as important as production. Brochures and handouts should be displayed at prominent points in the library. Posters may be placed next to printed indexes which are also available online. Printed materials should also be displayed in areas outside of the library where potential users normally congregate. This could be a departmental lounge or a grocery store bulletin board. If the library has the space, displays depicting the service can be a means of informing library users about the online search service.

Word-of-Mouth Publicity

While printed materials are the first idea that comes to mind when planning publicity, they are usually not the most effective. Word-of-mouth publicity is the most productive means of promoting the service. A satisfied user will generate new users from among personal acquaintances.

There are ways to stimulate this type of publicity. Identify key people who should be aware of the service. These may be faculty members at a university or managers in a business. They are individuals who direct the research or information activities of others. If necessary, go outside the library to meet with them. Offer to do a free search or send them a bibliography done online on a topic they are interested in. Vendors often offer free time on new databases or free time after training that a library can use for promotional purposes. Mailing brochures to key patrons is another means of alerting them to the service and is more effective than mass mailings.

Opinions formed by personalized contact last long and spread widely. All library staff members who come in contact with the public should be aware of the service and be able to answer basic questions or refer library patrons to an online searcher. Regular orientation meetings should be held for new staff to acquaint them with the service. The library staff attitude toward the service will have a strong influence on how the patrons they deal with regard it.

Demonstrations

After word-of-mouth publicity, demonstrations are probably the best means of informing users of the advantages of online searching. For publicity purposes, the library may set aside a day to run demonstrations or do free but limited searches. Demonstrations can be held in the library during hours of high use or by special invitation. Having preplanned searches on topics of known interest is a good idea. Searching on a professor's name to retrieve his publications or asking for search topics from the audience can demonstrate immediately the usefulness of the service.

Demonstrations may also be given outside the library to small, special-interest groups. If a portable terminal is available, searches may be run at faculty or administrative meetings. For university libraries, a short demonstration or talk can become a standard part of library instruction programs.

A prepared talk should always be a part of the demonstration. Point out the advantages of the service and how it can specifically aid the audience in their type of work or research. For the actual demonstration, a team approach works best. One librarian can run the search while another explains what is taking place. Remember, the size of a terminal display limits the number of people who can gather around it for a demonstration. Organize the demonstration accordingly by splitting up larger groups and positioning the terminal in the best location available.

When planning a demonstration downtime, or the unavailability of the system, should be taken into account. As a general rule, demonstrations should not be scheduled for peak vendor hours. Transparencies or slides showing the search strategies to be run may be used as a substitute for an actual search, especially when speaking to a large group.

Nonprint Material

Slides, film, or videotape may be used as part of a demonstration or as a separate publicity method. These should be short presentations (5-15 minutes) and, again, not overly technical. The videotape or slide show should point out what the search service does and how it can be accessed by the user. Some vendors and database producers have slide/script packages which you may purchase or borrow. These include: DIALOG Information Retrieval Service (12-minute color videotape and a slide/script presentation), Public Affairs Information Service (slide/sound presentation), ABI/INFORM (audio cassette), BIOSIS Previews (slide/script presentation), PSYCHINFO (slide/sound presentation), and PAI (audio cassette). They can be used as is or tailored to meet specific needs. Vendor promotional materials can also be a source of ideas for creating a slide presentation. Film is an especially effective means of communication, but it can be expensive and requires technical expertise.

Other Publicity Techniques

Public libraries may wish to consider professionally produced television or radio spots to advertise their search service. Local talk shows are another avenue for reaching a wide and varied audience. Newsletters and articles in the local or campus newspaper are also a means of reaching a wide audience. These must be repeated periodically, since their impact is short-lived.

Locating your search terminal in a prominent location is a simple but productive way to arouse interest in the search service. A prominent location for the terminal can have its drawbacks, since searchers may be interrupted while online. Even if the search terminal is not in public view, the search service should be in a convenient location and close to other public services.

Evaluating Publicity

All of these techniques may be used to promote online searching. Each should be evaluated before it is used. Display your slide presentation or show the services brochure to a small group of nonlibrarians. Be critical! Make sure the message being conveyed is clear.

After the publicity campaign has started, it will be necessary to evaluate the effectiveness of the different promotional materials and techniques. An easy way to determine how well the publicity is working is to ask before each search how the user found out about the service.

Conclusion

A few points should always be considered concerning the techniques discussed for advertising a computer search service. Do not mislead users about what the service can do. Users often have unrealistic expectations of what computer searching can achieve. The limitations of the service should be made as clear as the advantages. Remember, the purpose of the publicity is to make patrons aware of a service that *may* help them. If an online search is not what a patron needs, alternative library resources should be discussed.

The library should also be prepared for the results of its publicity campaign. Some attempt should be made at anticipating the demand the publicity will generate. Remember, good public opinion is the best promotion the service can have. If the quality of the search service drops because of large, unanticipated demand, poor public opinion will be the result. A poorly run search service will not be used, no matter how good the publicity is.

Finally, publicity is an ongoing project. Student bodies, employees, and neighborhoods all change. Keep the publicity current by changing its content, format or even techniques, as the situation demands.

Bibliography

Dragon, Andrea C. "Marketing the Library. *Wilson Library Bulletin* 53, no. 7 (March 1979): 498–502.

Edinger, Joyce A. "Marketing Library Services: Strategy for Survival." *College and Research Libraries* 41, no. 4 (July 1980): 328–32.

Ferguson, Douglas. "Marketing Online Services in the University." *Online* 1, no. 3 (July 1977): 15–23.

Hall, Vivian S. "Public Relations and the Librarian." *Southeastern Librarian* 28, no. 3 (Fall 1978): 177–80.

Kidd, J. C. "On-Line Bibliographic Services: Selected British Experiences." *College and Research Libraries* 38, no. 4 (July 1977): 285–90.

Schmidt, Janet A. "How to Promote Online Services to the People Who Count the Most ... Management ... End Users." *Online* 1, no. 1 (January 1977): 32–38.

Schmidt, Janet. "Online for an Online Public Relations Program." *Online* 2, no. 4 (October 1978): 47–50.

Sherman, Steve. *ABC's of Library Promotion*, 2nd ed. Metuchen, N.J.: Scarecrow, 1980.

REBECCA WHITAKER

The Impact of Online Search Services on Libraries

The impact of online search services on libraries is a vast topic. This discussion will focus on highlighting the variety of the effects an online search service can have on a library. The purpose of such a discussion is to make library professionals aware of these effects so that they will consider them in the planning process. Naturally, the extent to which a library experiences these effects may depend upon size, location, type, and other local factors.

A review of the literature on this topic is difficult since almost all literature dealing with online searching mentions some implication for the library situation. However, several substantial articles do exist. Hawkins's chapter on the management of search services in *The Library and Information Manager's Guide to Online Services* is a thorough presentation of the aspects of planning for the changes associated with online.[1] In *Librarians and Online Services*, Atherton and Christian stress the effects on staff and administrators, highlighting the need for an active role by the library administrator in the decision-making process.[2] Saffady's "Economics of Online Bibliographic Searching: Costs and Cost Justifications" surveys several effects but, as the title implies, stresses the cost justifications.[3] One of these justifications deals with the elimination of printed indexes once online service has been instituted.

The impact on interlibrary loan is one of the topics covered by White in his *Effects of Online Services on Other Library Functions*, a paper presented at the San Antonio Group of the Special Libraries Association. White also discusses library visibility and use by new clientele as a result of the introduction of online reference services.[4]

One landmark survey was sponsored by System Development Corporation (SDC). *The Impact of On-line Retrieval Services: A Survey of Users, 1974–75* by Wanger, Cuadra, and Fishburn surveyed library administrators' and searchers' perceptions and attitudes about the impact of online on their libraries.[5] This study was done when libraries just began to offer online searching services. Now, after five years and twice the experience, a follow-up study to see if perceptions have changed might be interesting. For example, has the original survey group undergone any attitudinal changes? More importantly though, have newly inaugurated online search services learned from the experience of the first group, or have they made the same errors?

Effects during the Planning Period

Even before the first search is performed, the decision to initiate an online search service can affect the library and its staff. The morale of the staff and its attitude toward this new service can make or break the project. Kusack, in his discussion "Integration of On-Line Reference Service," observed that a negative attitude on the part of the librarian not only fails to instill confidence in online services, but also does nothing for the image of traditional reference services.[6] As with any automated service, staff members need to be thoroughly introduced to the concepts and capabilities of database searching in advance so that they are aware that changes will occur.

In order to initiate service, the library must accept the concept of continuing education, because, typically, searchers are selected from the existing reference department staff. Firschein, Summit, and Mick, reporting about the public library DIALIB experiment, state that traditionally trained librarians with experience in using manual search tools and with communications and reference interview skills can easily transfer these skills to online database searching.[7] To make this transfer of skills, however, training and continuing education are essential.

New policies and procedures may be necessary to accommodate online searching. Online search services often operate by appointment rather than by casual drop-in. This is contrary to traditional reference service, for which no appointment is needed. Librarians may also find themselves doing some tasks which they may identify as clerical: logging searches, completing bills, distributing search results, and collecting money. In some libraries these tasks are done by support staff. A written statement detailing specific policies and modes of operation can give everyone a better understanding of what is expected.

Online searching provides many libraries with the opportunity to extend their philosophy of reference service. Wanger, Cuadra, and Fishburn found that the availability of online searching allowed for the introduction of literature-searching services. Sixty percent of the educational libraries not previously offering custom-made bibliographies were able to do so with the initiation of online searching. In special libraries the figure was somewhat lower, probably because these libraries provided some type of literature searching before online service was available.[8]

In school and academic libraries, library instruction is often an integral part of the reference service. Some librarians may see online search services as contrary to this philosophy, since currently the actual online search is usually done by the librarian and may be seen to obviate the library user's need to learn how to use reference tools. However, as Dreifuss points out, motivation of students is critical for a successful library instruction program. Online searching can act as a motivating force and the presearch interview process can provide an opportunity for point-of-interest library instruction.[9]

The new online search service needs to be integrated into the organization of the library. The decision to place it in the realm of the reference department, to establish a separate entity, or to create some hybrid pattern will affect staffing. For example, if the responsibility for the services lies outside the reference department but reference staff are doing the searching, the staff may feel confused about working for two bosses. Also, the head of reference may not be involved in an activity that can consume much reference staff time.

The library administration needs to consider also the effects on the budget. Will existing funds be re-allocated (always a difficult task), or will new monies be acquired? The second option may involve exploring unfamiliar funding sources such as foundations, local businesses, and civic organizations.

Related to budgeting and funding is of course the philosophical question of fee versus free. There is now ample literature on this topic. Drake's recent compilation of articles reviews the issues and presents opposing views and case studies.[10] The bottom line seems to be that each library must assess its local situation: What is your philosophy of service? What are your service patterns? What alternatives for funding exist? Who are your clientele? and so forth. Based on the local situation, a decision must then be made. The final report of the DIALIB public library project observes that the levying of fees for online services had little or no adverse effect on users' attitudes. One problem librarians faced, however, was the philosophical concept of charging for service.[11]

Effects after Initiation of the Service

So far the changes discussed occur during the planning phase of an online search service. Library personnel need to be aware that more changes are still to come. Once the search service has been inaugurated and the use of the specialized subject indexes available online increases, several departments in the library may feel the effects. The most obvious is probably the reference department.

New procedures and policies, new definitions of reference service, and new staff responsibilities have already been mentioned as effects on the reference department. All of this may mean increased work loads or reassignment of staff as the searching activity increases. Unfortunately for searchers, the majority of the managers surveyed by Wanger et al. did not perceive that increased work load necessitated adding additional staff.[12] However, productivity apparently also increases with the introduction of online searching. The Systems Development Corporation (SDC) study asked managers to respond to the following statements: "The productivity of staff is greatly increased by the use of on-line services" and "On-line searching allows the information specialist or librarian to spend his/her time more productively." Over 75 percent of the library managers agreed with both statements. Perhaps this perceived increase in productivity is presumed to help offset this increased work load.[13]

The generation of more meaningful statistics about a reference service has a positive effect on reference departments. The record keeping usually recommended for online search services, specifically search log sheets, can provide hard data on the volume of searches performed, databases accessed, types and numbers of patrons served, and the time involved per search. These statistics can be more reliable than the familiar "hash-mark" method of tallying reference encounters.

The enhancement of the library profession after the introduction of online searching is, perhaps, one of the more important effects. The self-esteem of librarians is raised as they participate as active and willing partners in the information retrieval process. New job titles such as "search analyst," "search consultant," "information retrieval specialist," and "information scientist" reflect this new self-esteem. Atherton and Christian discuss this point and emphasize that new technology and skills provide better service in the eyes of the patron.[14]

The additional access points, the ease of searching complex logical relationships to a topic, and the short time needed to complete a retrospective search of several years or databases, monographs, journal articles, documents, and reports that were difficult to identify manually are more readily identified online. Often an abstract accompanies the citation so that the user can select pertinent articles with more assurance. Moreover, with less time involved in the actual search process, the requester may have time for a more complete reading of the literature. Furthermore, unlike a manual search which takes days or weeks and in which only a few items may be located at any one time, the online search can immediately provide many potentially relevant citations. Document delivery now becomes the issue.

Therefore, the interlibrary loan department commonly is cited as an area in which online searching has a great effect. White's review of online searching impacts covers this area in detail. He concludes that while the number of interlibrary loan transactions tends to grow, the extent of this growth may depend upon the location, type, and size of the library as well as the subjects searched and the make-up of the clientele.[15] Automation of interlibrary loan procedures can help counter the increased work load created by online searches. The advent of interlibrary loan fees and online ordering of documents are other options libraries have employed to relieve pressures on interlibrary loan departments. Furthermore, while the introduction of an online search service often increases interlibrary loan transactions, the resulting requests can also be used to identify significant gaps in a library's collection and therefore to aid in collection development.

The circulation of a library's own collection and the in-house use of materials can also increase. Perhaps the most telling development can be the increased use of the journal collection. Bibliographical databases typically include a large percentage of journal article citations. Online searching, as has been noted, facilitates signaling out useful citations and can, therefore, increase journal usage. Unfortunately, Atherton and Christian point out, the greater use of materials may mean greater abuse also.[16] Two seldom mentioned trickle-down effects can be the rise in demand for photocopying and the increased volume of shelving to be done.

Another effect on collection development has been the tendency by some administrators to see online as a rationale for canceling print-version subscriptions. Saffady points out there is a break-even point for every printed index. The task is to determine what that point is, based on the subscription cost and the number of manual uses, compared to the cost of searching the online counterpart for the same number of users. Saffady's formula for computing break-even points takes into consideration subscription prices, technical processing and overhead costs, and all the associated online costs.[17] However, this formula ignores other factors such as the need for multiple user access, appropriateness of online searching for the topic, and completeness of the online version, for example, the presence or absence of abstracts.

In general, Lancaster and Goldhor found that in a survey of university libraries, online searching capabilities have so far had only a minor effect on print subscriptions. There does seem to be evidence, though, that some libraries are considering a re-evaluation of their selection policies concerning print subscriptions versus online access.[18] Librarians, furthermore, may find themselves in a Catch 22 situation as publishers justify increases in subscription prices because of cancellations and fewer new subscriptions, while librarians decide they cannot afford the resulting higher prices. The only alternative may be online access.

Just as the professional staff is affected by the introduction of an online search service, so is the support staff. Clerical staff in the reference and accounting departments may be involved in record keeping, billing procedures, and appointment scheduling. Circulation staff may be responsible for the distribution of search results and the collection of fees. Of course, if support staff does not exist in these areas, librarians may have to assume these nonprofessional duties.

Finally, there is also an effect on library users. The previous discussion has already alluded to changes in users' perceptions of the library profession and the increased use of library collections and other services. Hawkins and White are among the authors noting that new patrons who have not been regular library users can be attracted to the library because of online searching. The search requests of these new patrons may be more technically oriented and require quick responses.[19] On the other hand, some patrons may be attracted by the technology itself.

Library patrons may need to learn new skills. Users accustomed to doing their own manual searching, for example, will need to adjust to changes in the search process. One significant change involves how the results of an online search are used. During a manual search, the searcher can easily screen out irrelevant citations as they are found. However, the result of an online search is often a large collection of both relevant and nonrelevant citations that must then be screened.

In summary, this discussion has presented an overview of the implications of online search services for libraries. Unlike some automated technical services, in which most effects are hidden from the public, online database searching is a public service which has overt effects on the library organization, operations, staff, and the library users. The overriding impact is the necessity to re-assess a library's philosophies of service and of access to material, and thus to address the fee versus free question.

An online search service enhances the image of the library profession and, by extension, the reputation of the library. The visibility of an active, well-managed online search service can promote the library as an innovative organization keeping abreast of technological enhancements in information management and access. Because of these seemingly vast implications, awareness of these effects not only offers help in the planning and implementation of a successful service, but demands the effective involvement of the library administrator in the decision-making process.

Notes

1. Donald T. Hawkins, "Management of an Online Information Retrieval Service," in *The Library and Information Manager's Guide to Online Services*, Ryan E. Hoover, ed. (White Plains, N.Y.: Knowledge Industry Publications, 1980), pp. 97–125.
2. Pauline Atherton, and Roger K. Christian, *Librarians and Online Services* (White Plains, N.Y.: Knowledge Industry Publications, 1977).
3. William Saffady, "The Economics of Online Bibliographic Searching: Costs and Cost Justifications," *Library Technology Reports* 15 (September–October 1979): 567–653.
4. Phillip M. White, *Effects of Online Services on Other Library Functions*. Presented at the San Antonio Group Special Libraries Association, of Texas Health Science Center, November 7, 1980 (Arlington, Va.: ERIC Document Reproduction Service, ED 195 271, 1980).
5. Judith Wanger, Carlos Cuadra, and Mary Fishburn, *Impact of On-Line Retrieval: A Survey of Users, 1975–75* (Santa Monica, Calif.: System Corporation, 1976).
6. James M. Kusack, "Introduction of On-Line Reference Service," *RQ* 19 (Fall 1979): 64–69.
7. Oscar Firschein, Roger K. Summit, and Colin K. Mick, "Planning for On-Line Search in the Public Library," in *User Fees: A Practical Perspective*, ed. Miriam A. Drake (Littleton, Colo.: Libraries Unlimited, 1981), pp. 85–93.
8. Wanger et al., *Impact of On-Line Retrieval*.
9. Richard A. Dreifuss, "Library Instruction in the Database Searching Context," *RQ* 21 (Spring 1982): 233–38.
10. Miriam A. Drake, ed., *User Fees: A Practical Perspective* (Littleton, Colo.: Libraries Unlimited, 1981).
11. Roger Summit and O. Dirschein, *Final Report: Investigation of the Public Library as a Linking Agent to Major Scientific, Educational, Social and Environmental Data Bases* (Palo Alto, Calif.: Information Systems Programs, Lockheed Palo Alto Research Laboratory, 1977).
12. Wanger et al., *Impact of On-Line Service*.
13. Ibid.
14. Atherton and Christian, *Librarians and Online Services*.
15. White, *Effects of Online Services*.
16. Atherton and Christian, *Librarians and Online Services*.
17. Saffady, "The Economics of Online Bibliographic Searching."
18. F. W. Lancaster and Herbert Goldhor, "The Impact of Online Services on Subscriptions to Printed Publications," *Online Review* 5 (August 1981): 301–11.
19. Hawkins, "Management of an Online Information Retrieval Service"; White, *Effects of Online Services*.

APPENDIX

An Introduction to Online Searching: A Suggested Outline

Prepared by MARS EDUCATION AND TRAINING
OF SEARCH ANALYSTS COMMITTEE

Introduction

While online information retrieval services are becoming more prevalent in all types of libraries, there is still a significant number of librarians who have had no exposure to this relatively new library service. Therefore, those librarians with online searching experience are frequently asked to present introductory workshops or demonstrations of online searching to their colleagues or library users. The following outline is designed to aid librarians when they plan those types of introductory sessions. Furthermore, the outline can also be used as a checklist or a review aid by someone familiar with online searching.

The purpose of a workshop based on this outline should be to introduce and explain the basic concepts of computer searching and to show its applications in library reference work. Such a workshop is intended for an audience with no previous online experience or exposure. It can serve as an educational experience for a total library. In addition, this workshop can serve as a foundation for those who subsequently go on for training as search analysts.

As its title indicates, the outline is suggested as an approach to presenting an introductory workshop. It can be used as a planning tool for designing various types of introductory workshops. The outline consists of elements to be included and lists relevant points of information which should be covered for each element. For some audiences, this information may need to be expanded (e.g., definitions of libraries and online terminology for a library patron audience) or specially tailored (e.g., examples of education databases for a school librarian audience). The elements themselves may be rearranged or eliminated depending on the focus of the workshop, the needs of the audience, and/or time constraints. One important element which is not men-

MARS Education and Training of Search Analysts Committee, Reference and Adult Services Division, American Library Association: Becki Whitaker (chair, 1982), Greg Byerly (chair, 1981), Norene Allen, Stephen Harter, Doug Jones, Maureen Pastine, C. Patricia Riesenman, Marilyn Cabonell, M. Virginia Jackson, Carol Tobin.

tioned in the outline is the use of online demonstrations. Since concepts are introduced which may be confusing, if not mystifying, an online demonstration should always be included in an introductory workshop on online searching. Most search service vendors are willing to provide online time for demonstrations to groups of potential users. However, policies vary from vendor to vendor so presenters should individually investigate the possibility of free time with prospective search services. The marketing department or regional sales representative of a vendor should be approached concerning demonstration time.

This outline is designated to be compatible in purpose, audience level, and other workshop components with those recommended for an independent introductory workshop in "Online Training Sessions: Suggested Guidelines," *RQ*, 20 (Summer 1981): 353–57. The material covered by the outline presents an overview of the basic concepts of online searching and applications for libraries. It is assumed that the audience has no online experience or previous exposure to online searching. The presenter should be an experienced searcher (there may be more than one) who has ample library experience plus familiarity with a variety of databases and vendors. A lecture, illustrated with slides or overhead transparencies, and online demonstrations are integral parts of an introductory workshop which covers new concepts and terminology. The guidelines also recommend the length of the session (½ or 1 day) and audience size (30–50 people), but local factors in part dictate which outline elements are covered, which omitted, which are stressed and which briefly mentioned. Anyone planning a workshop from the outline is encouraged to consult the guidelines as a planning aid.

One program which was developed from this outline was the MARS Program at the 1982 ALA Annual Meeting in Philadelphia. For this program, the elements of the outline were divided into two categories: introduction of online services and the management of online services. Two half-day sessions were then structured covering these subjects. The introductory session was designed for libraries with little or no experience in online searching. It included a historical perspective, advantages and disadvantages of using online, basic equipment requirements, applications of online, and an overview of database procedures and vendors. For librarians who need to know about organizing and operating online search services, the management session covered financial considerations, methods of funding, mechanics of searching and training, and impacts on the library. Since the outline was not intended to cover all the aspects of the management of an online service, discussions of policy formation and forms were added to the management session by the 1982 MARS Program Committee.

Due to the anticipated large audience of 500 or more persons, a workshop atmosphere was impossible. The important online demonstration was also eliminated because of the cost of equipment for such a large group demonstration. Rather, the program committee selected speakers who presented papers that addressed specific elements of the outline. These papers are collected in this volume and are valuable resource materials for planners of introductory online workshops. The MARS ETSA Committee recommends that these materials be used as a starting point in gathering background material for an introduction to online workshop.

An Introduction to Online Searching: A Suggested Outline

I. General introduction and definition of terms
   A. *Online Searching* is a means of finding desired information, normally bibliographic references (although this is changing rapidly) by using a computer in an interactive mode.
   B. *Database* is a collection of information, called machine-readable records, often corresponding to a printed index.
II. Brief historical background
   A. Batch processing to computer time-sharing
   B. Telecommunications networks and remote terminals
   C. Development of national search services
      1. DIALOG, SDC, and National Library of Medicine (1960s)
      2. BRS and others (1970s)
III. Advantages of online searching
   A. Search years of indexes quickly to find references to very specific topics
   B. Find information on a topic not approachable through a printed subject index
   C. Are typically more current than printed indexes
   D. Offer access to many indexes not owned in print by the library
   E. Provide original bibliographies made just for a particular topic or patron
   F. Provide a cost-effective alternative to many kinds of library research
   G. Provide selective dissemination of information services
   H. Offer access to some indexes that are available only online
IV. Limitations of online searching
   A. Normally cannot provide direct answers to questions (although this is changing rapidly)
   B. Results not limited to references located in the library
   C. Cannot write articles or papers
   D. Normally does not provide references to older journals or other literature
V. Initial considerations for an online search
   A. Complicated topics
   B. Topics which are not approachable through a printed index
   C. Topics for which comprehensive coverage is desired
   D. Questions for which a manual search would be time-consuming
   E. Very current topics
   F. Some nonbibliographical or statistical questions

G. Bibliographic verification
H. Topics for which printed indexes are not locally available
VI. Applications in a library
A. *Types of general uses* (These will vary depending on the library.)
1. *Bibliographies:* Quickly and comprehensively provide bibliographies, often annotated, on a wide variety of subjects for both patrons and librarians
2. Ready-reference Uses: Directly provide answers to some types of questions, especially those requiring either very current sources of information or bibliographic verification
3. Nonbibliographical references: Primarily business data and statistics
VII. Databases and database producers (More than 100 unique databases are currently available from the major search services.)
A. Types of databases
1. Bibliographic (e.g., Sociological Abstracts database)
2. Directory (e.g., Encyclopedia of Associations database)
3. Numerical/statistical (e.g., PREDICASTS)
4. Full text (e.g., NEXIS)
B. Types of database producers
1. Associations and societies (e.g., MLA)
2. Government (e.g., NTIS)
3. For profit (e.g., ISI)
C. Comparison of printed indexes to corresponding databases
1. Directly equivalent (e.g., ERIC)
2. Enhanced online (e.g., PSYCH INFO or NIMIS)
3. No print equivalent (e.g., ABI/INFORM)
4. Microfilm equivalent (e.g., Magazine Index database)
VIII. Database vendors and search services
A. Representative search services (database vendors)
1. Bibliographic Retrieval Services (BRS)
2. DIALOG Information Services
3. National Library of Medicine (NLM)
4. New York Times INFOBANK
5. System Development Corporation (SDC)
6. Many others
B. Definition of the relationship of database producers, search services, telecommunications
C. Availability of search services
1. Individual contracts with each search service
2. Consortia, regional networks, user groups
3. State agencies
IX. Financial considerations
A. Direct costs
1. Online connect time

  2. Telecommunications connect time
  3. Print charges
 B. Indirect costs
  1. Terminal, modem, and supplies
  2. Database aids and manuals
  3. Promotion and publicity costs
  4. Staff training
  5. Staff time
  6. General overhead
  7. Clerical support
X. Methods of funding
 A. Sources of funding
  1. New funds
  2. Grants or special appropriations
  3. Reallocation of existing funds
  4. User fees
 B. Types of user fees
  1. Partial charge of direct costs
  2. Full recovery of direct costs
  3. Full recovery of direct costs, plus some additional indirect costs
  4. Higher fees for nonaffiliated users
XI. Contract options and arrangements
 A. Use only
 B. Prepaid subscription
 C. Discounted rate for guaranteed number of online hours
 D. Group contracts
XII. Mechanics of searching
 A. Mechanical needs
  1. Terminal
   a) Print vs. CRT; low vs. high speed; portable vs. stationary; lease vs. buy; methods of purchase, etc.
  2. Telecommunications
   a) Telenet vs. Tymnet vs. Uninet; direct dial, etc.
 B. Search process
  1. Information interview
   a) Online searching is very much a human process and requires a comprehensive reference interview before conducting the search.
  2. Database selection
  3. Search options
   a) Search by descriptors (subject headings)
   b) Search by words in title or abstract
   c) Search in other fields (language, date, type of publication, etc.)
  4. Search strategies

            a) Boolean logic—AND, OR, NOT
            b) Truncation
            c) Free text
         5. Online search
            a) Online search is conducted with the computer and the strategy is modified, as needed. Each database is searched individually and many searches may be conducted on more than one database.
         6. Postsearch
            a) Results of the search are delivered and explained.
            b) For lengthy bibliographies, results are printed offline, mailed to the library, and normally received within 3–5 working days.
            c) Immediate results can, however, be printed online when the actual search is conducted.
XIII. Librarians as search analysts (Reference librarians with necessary subject background and appropriate skills and personality traits are usually selected and trained to do the actual searching. However, all reference librarians are involved in determining what topics to recommend for a computer search and in initially explaining the service to patrons. In many cases both the referring librarian and the librarian trained in searching will jointly prepare the search strategy.)
   A. Necessary attributes
      1. Reference ability
      2. Logical skills
      3. Willingness to make presentations and perform demonstrations
      4. Willingness to practice and keep current in the field

      Note: Typing and terminal proficiency are much less important than the librarian's interviewing skills and ability to devise an intelligent search strategy and to modify that strategy according to intermediate feedback. No prior knowledge or understanding of computers is necessary.
XIV. Education and training of librarians and search analysts
   A. Training sources
      1. Self-taught
      2. Library schools
      3. In-house training
      4. Search service (vendor) training sessions
      5. Database supplier (producer) training sessions
      6. Independent workshops or clinics
   B. Continuing education sources
      1. All of the above
      2. Online users' groups
      3. Professional organizations (e.g., ALA/RASD/MARS)
      4. Consortia and regional and local networks

Note: Formal and ongoing training of computer searchers is essential, but equally important is the need to educate all members of the library staff concerning online reference services and to increase their familiarity with the new techniques.
XV. Organization within library
- A. Centralized or dispersed authority
- B. Separate unit or integrated into reference unit
- C. Policies and procedures (e.g., hours of operation)
- D. Physical facilities and environment (e.g., location of service)

XVI. Impact on the library
- A. Effects on various departments
    1. Administration
    2. Circulation
    3. Acquisitions
- B. Effects on interlibrary loan
- C. Effects on photocopying
- D. Effects on the reputation of the library and librarians

XVII. Future trends
- A. Number and kind of databases and search services
- B. Cost trends
- C. End users
- D. Service growth

BIBLIOGRAPHIES

Online Reference Service: How to Begin

Edited by EMELIE J. SHRODER

The materials in this bibliography were selected and are recommended by the RASD MARS Use of Machine-Assisted Reference in Public Libraries Committee members: Rachael Gade, Jane Light, Bernard Pasqualini, Gene Rollins, Tina Roose, Richard Waters, and Emelie Shroder, chair.

Some of the references are annotated when necessary to clarify subject matter. It is suggested that at least one of the books and two or three of the other references from the list of general materials should be read first for an overview, and that the other subjects be investigated as needed for more in-depth information.

Though this bibliography was prepared by librarians in public libraries, a very large percentage of the material relates equally well to a university or special library setting. The articles which discuss the fee-vs.-free charge issue are essentially the only exceptions for the special library.

General References

Atherton, Pauline, and R. W. Christian. *Librarians and Online Services.* White Plains, N.Y.: Knowledge Industry Publications, 1977.
 Includes chapters on start-up, finance, impact on library staff and administrators, marketing, and promotion of online services.
Bourne, C. P. "Online Systems: History, Technology, and Economics." *American Society for Information Science Journal* 31 (May 1980): 155–60.
 Traces the evolution of online bibliographic databases, vendors, searching, hardware, and pricing.
Chen, Ching-Chih, and Susanna Schweizer. *Online Bibliographic Searching: A Learning Manual.* New York: Neal-Schuman, 1981.
 A beginner's guide to the skills necessary for retrieval of information from computerized bibliographic files. Designed to serve as a self-instruction manual using DIALOG as an example throughout the text. The authors suggest that the "manual is intended to serve as a tool for self-instruction as a part of an ongoing, hands-on learning experience." There are also chapters

on search service management, search techniques, reference interviews, database vendors, and a glossary.

Fenichel, Carol H., and Thomas H. Hogan. *Online Searching: A Primer.* Marlton, N.J.: Learned Information, 1981.

Assumes no previous knowledge of the subject and contains a step-by-step explanation of how to start an online service. Includes chapters on types of databases, vendors, terminals selection, basics of search techniques, training, reference interviews, and service charges.

Hawkins, Donald T. *Online Information Retrieval Bibliography 1964–1979.* White Plains, N.Y.: Learned Information, 1980.

Reprint of the original bibliography and its first three updates which appeared in the 1978 supplement; March 1979; and March 1980 issues of *Online Review*. The fourth update is in the April 1981 issue of the journal (pp. 139–82). Use this if further articles on any subject are needed.

Hoover, Ryan E., ed. *Library and Information Managers' Guide to Online Services.* White Plains, N.Y.: Knowledge Industry Publications, 1980.

Includes sections on types of databases available, producers and vendors of databases, management of online information retrieval services, promotion, training, mechanics of searching, and online user groups, but does not deal in detail with hardware.

Johnson, Robert A. "Planning for Online Searching at San Jose: A Design for Public Libraries of the 1980's." *Science and Technology Libraries* 1:117-132 (Fall 1980).

Katz, Bill, ed. *Reference and Online Services Handbook: Guidelines, Policies and Procedures for Libraries.* New York: Neal Schuman, 1982.

Includes sections on interviewing procedures, search results, document delivery, patron evaluation and ready-referenec use. This book also includes a section of excerpts from online policy statements from academic, public, and special libraries.

Keenan, Stethe, Nick Moore, and Anthony Oulton. "Online Information Services in Public Libraries." *Journal of Librarianship* 13 (January 1981): 9–24.

Kusack, James M. "Integration of On-line Reference Services." *RQ* 19 (Fall 1979): 64–68.

———. "Online Reference Service in Public Libraries." *RQ* 18 (Summer 1979): 331–34.

McClure, Charles R. "A Planning Primer for Online Reference Service in a Public Library." *Online* 4 (April 1980): 57–65.

Meadows, Charles T., and Pauline Atherton Cochrane. *Basics of Online Searching.* New York: John Wiley & Sons, Inc., 1981.

The purpose of the book is to explain, for those with little or no experience, how to search a database with the aid of a computer. According to the authors, the book can be used as a self-instruction manual, but "it must be coupled with practice and the use of prepared exercises and search service users' manuals." There is a useful chapter on terminals and networks.

Raedeke, Amy. "Machine Assisted Reference Service in a Public Library: A One-Month Test Period." *Online* 2 (October 1978): 56–59.

Describes an experiment at Minneapolis Public Library where free online searches were promoted by the reference staff. Presents statistics on types of databases used, cost, and relevance of citations retrieved.

Waters, Richard L., and Jane Mann. "Online Search Service at the Dallas Public Library." *Science and Technology Libraries* 1 (Fall 1980): 109–15.

Financial Aspects

For information on finance and costs, also check the General References section. For current trends on charging, the book by Mary Jo Lynch, listed below, *Financing Online Search Services in Publicly Supported Libraries*, is very important.

Blake, F. M., and E. L. Perlmutter. "Rush to User Fees: Alternative Proposals." *Library Journal* 102 (October 1, 1977): 2005–8.

Discusses the free-vs-fee question and offers alternatives to user fees.

Boyer, Bert R., and Edward I. Gillen. "Is It More Cost-Effective to Print on or Offline?" *RQ* 21 (Winter 1981): 117–20.

Crawford, P. J., and J. A. Thompson. "Free Online Searches Are Feasible." *Library Journal* 104 (April 1, 1979): 793–95.

Describes how the library at California State College, Stanislaus, integrates online services into its general reference service.

Drinan, Helen. "Financial Management of Online Services—A How-to Guide." *Online* 3 (October 1979): 5–23.

Elman, S. A. "Cost Comparison of Manual and On-line Computer Literature Searching." *Special Libraries* 66 (January 1975): 12–18.

Reports the results of a study which concludes that thorough literature searches are far more expensive when performed manually.

Flynn, T., et al. "Cost Effectiveness Comparison on Online and Manual Bibliographic Information Retrieval." *Journal of Information Science* 1 (May 1979): 77–84.

Knapp, Sara D., and James C. Schmidt. "Budgeting to Provide Computer-Based Reference Services." *Journal of Academic Librarianship* 5 (March 1979): 9–13.

Koch, Jean E. "A Review of the Costs and Cost-Effectiveness of Online Bibliographic Searching." *Reference Services Review* 10 (January–March 1982): 59–64.

Kranich, Nancy. "Fees for Library Service: They Are Not Inevitable." *Library Journal* 105 (May 1, 1980): 1048–51.

Linford, John. "To Charge or Not to Charge: A Rationale." *Library Journal* 102 (October 1, 1977): 2009–10.

Article attempts to provide a decision-making framework for the question of whether to charge for online searches.

Lynch, Mary Jo. *Financing Online Search Services in Publicly Supported Libraries: The Report of an ALA Survey.* Chicago: American Library Association, 1981.
  Trends in policies in the financing and charging for online searching.
———. "Libraries Embrace Online Search Fees." *American Libraries* 14 (March 1982): 174.
  A synopsis of the ALA report listed above.
Saffady, William. "The Economics of Online Searching: Costs and Cost Justification." *Library Technology Reports* 15 (September–October 1979): 567–654.
Waldhart, T. J., and T. Bellardo. "User Fees in Publicly Funded Libraries." In *Advances in Librarianship* 9 (1979): 31–61. New York: Academic Press, Inc., 1979.
Watson, Peter. "The Dilemma of Fees for Service: Issues and Action for Librarians." *ALA Yearbook* (1978): xv–xxii. Chicago: American Library Association, 1978.

Equipment and Communication Considerations

Auerbach and Datapro have services that rate and describe terminals and other kinds of equipment. Probably easier to find and use are individual issues of *Library Technology Reports* and *Computer Equipment Review.*

All about Alphanumeric Display Terminals. Reports on Data Communication. Delran, N.J.: Datapro Research Corp., 1982.
All about Teleprinter Terminals. Reports on Data Communication. Delran, N.J.: Datapro Research Corp., 1982.
Becker, Joseph. "Printer Terminals for Libraries." *Library Technology Reports* 16 (May–June 1980): 231–90.
  Has occasionally reviewed equipment that can be used for online searching.
Computer Terminal. Auerbach Computer Technology Reports. Pennsauken, N.J.: Auerbach Publishers, 1981.
Crawford, Walt. "CRT Terminal Checklist." *Journal of Library Automation* 13/1 (March 1980): 36–45.
  Following an evaluation of the institution's needs, three principal steps are described: (1) assembling information from advertisements and literature relative to display characteristics, keyboard, display speeds, power requirements, special features, price, and availability; (2) direct observation and use, paying attention to additional display characteristics, keyboard characteristics, environmental characteristics, and maintenance and durability; and (3) talking to other users of the terminal.
Data Communications Equipment. Auerbach Computer Technology Reports. Pennsauken, N.J.: Auerbach Publishers, 1981.
Online Terminal Guide and Directory. Terminal Guide & Directory. Weston, Conn.: Online, Inc., 1979–80.
Pemberton, Jeffery K. "Should Your Next Terminal Be a Computer?" *Database* 4 (September 1981): 4–6.

Suggests that microcomputers should be seriously considered as an alternative to a video terminal and printer. Includes cost estimates and a list of what to look for when shopping for your first microcomputer.

Radwin, Mark. "Choosing a Terminal." *Online* 1 (January 1977): 11–17 and 1 (April 1977): 61–63.

Saffady, William. "Selecting Interactive Terminals for Library Applications." *Library Computer Equipment Review* 1 (1979).

This semi-annual periodical has regular reviews of computer terminals and other equipment needs. The title was changed for Fall 1981 (v. 3, no. 1) to *Computer Equipment Review*.

Comparing and Rating Databases and Database Systems

Several periodicals regularly review, compare, and rate databases and database systems: *Database, Byte, RQ, Online Review, Reference Services Review,* and *Online*. The following are examples of the kinds of articles available.

Caldwell, Jane, and C. Ellington. "A Comparison of Overlap: ERIC and Psychological Abstracts." *Database* 2 (1979): 62–67.

Kritchman, Albert. "Command Language Ease of Use: A Comparison of DIALOG and ORBIT." *Online Review* 5 (June 1981): 227–40.

Lieberman, Vilma M., and Bernard Pasqualini. "The Information Bank II." *RQ* 21 (Fall 1981): 83.

Miastkowski, Stan. "Information Unlimited: The Dialog Information Retrieval Service." *Byte* (June 1981): 88–108.

Description of DIALOG; includes an overview of searching and briefly discusses costs.

Prendergast, Kathleen. "National Criminal Justice Reference Service." *RQ* 21 (Fall 1981): 85.

Swanson, Rowena. "Probing Private Files." *Database* 3 (June 1980): 70–75.

A comparison of SDC, DIALOG, and BRS.

Advertising and Promotion

Many of the books under General References also contain good information about marketing your service.

Antony, Arthur, and Eugene Graziano. "Impacts of The Promotion of Online Services in Academic Libraries." Bethesda, Md.: ERIC Document Reproduction Service, 1980. ED 197739.

Paper presented at Online 80 Meeting, San Francisco, October 1980. Survey of promotional policies and practices for online services in 101 research libraries addressing the effectiveness and the impact of those practices. Respondents report as most successful word-of-mouth promotion, supplemented by mailings to specific audiences. Includes a literature survey and a sample promotional brochure.

Crane, Nancy B., and David M. Pilachowski. "Introducing Online Bibliographic Service to Its Users: The Online Presentation." *Online* 2 (October 1978): 30–29.

Discusses advantages and disadvantages of online searching.

Ferguson, Douglas. "On-Line Services in the University." *Online* 1 (July 1977): 15–23.

Stresses need for systematic, paced marketing effort. Includes illustrations of promotional materials found effective in academic libraries. Recommends that the focus of the marketing thrust be relevant to the user.

Quinn, Karen Takle. "Promoting Your Information Services by Education Managers and Users." In *Online '81 Conference Proceedings*, pp. 361–63. Weston, Conn.: Online, Inc., 1981.

Offers specific suggestions for selling online information services to managers and users.

Seba, Doug. "How to Do a Demo." *Online* 2 (October 1978): 53–55.

Smith, Patricia K. "Marketing Online Services, Part 1." *Online* 4 (January 1980): 60–62.

⸺⸺⸺. "Marketing Online Services, Part 2." *Online* 4 (April 1980): 68–69.

Part 1 introduces fundamental marketing concepts and stresses awareness of the market's perception of the product. Part 2 offers "strategies for the packaging of online services" including specifics of product design and pricing.

Wilkens, Carol D. "Online Bibliographic Retrieval in Public Libraries: The Windy City Experience." In *Online '81 Conference Proceedings*, pp. 73–78 (esp. 75–76). Weston, Conn.: Online, Inc., 1981.

Suggests providing for general staff orientations early in the promotional program. Recommends keeping library limitations in mind. Lists specific tactics. Refers to the literature.

Training of Searchers

Searchers need training, preferably from an online service vendor, to do more efficient searching. An opportunity to observe an experienced searcher conduct some online searches is useful. In addition, studying an online user manual from a vendor or one such as Chen, *Online Bibliographic Searching: A Learning Manual*, mentioned below, is desirable for a beginning searcher. Eventually, after the searcher has conducted many searches, training sessions focusing on a specific database subject, business, for example, or an advanced training session from a vendor are excellent for bolstering search skill and efficiency.

Bourne, C.P., and J. Robinson. "Education and Training for Computer-Based Reference Services: Review of Training Efforts to Date." *Journal of the American Society for Information Science* 31 (January 1980): 25–35.

Chen, Ching-Chih. "Online Bibliographic Searching: A Learning Manual."

See complete citation under General References. Very strong chapters on searching and training mechanisms.

Kuroki, Kristyn. "Online Regional and On-Site Training Opportunities in Lockheed, SDC, and BRS systems and Their databases." *Online* 3 (July 1979): 37–49.

Swanson, Rowena. "An Assessment of Online Instruction Methodologies." *Online* 6 (January 1982): 38–54.

Discusses the results of a survey administered to former students which compares nine online instruction programs offered by various organizations. All online instruction program characteristics and questionnaires used are included. Suggests tools to improve learning.

Tedd, Lucy A. "Teaching Aids Developed and Used for Education and Training for Online Searching." *Online Review* 5 (June 1981): 205–16.

The English viewpoint.

Tenopir, Carol. "An In-House Training Program for Online Searchers." *Online* 6 (May 1982): 20–26.

Van Camp, Ann. "Effective Search Analysts." *Online* 3 (April 1979).

Description of the qualities required in a good online searcher.

User Aids

There are user aids and/or thesauruses for each individual database. These are essential during the reference interview in order to develop fine-tuned and cost-effective search strategies. The user aids are available from individual database producers and from the database vendors. Addresses for ordering are found in the database directories in the next section of the bibliography.

Conger, Lucinda D. "Online Command Chart," 2nd ed. Available from Online, Inc., 11 Tannery Lane, Weston, Conn. 06883.

All orders must be prepaid. $10 for the first chart, $6 for each additional chart.

Unruh, Betty. "Database User Aids and Materials—A Study." *Online Review* 5 (February 1981): 7–24.

Good overview of search aids available.

Database Directories

These directories provide descriptions of the individual databases available from database vendors such as DIALOG, BRS, SDC, and many others.

Directory of Online Databases. Santa Monica, Calif.: Cuadra Associates, 1979.

Published quarterly. Good descriptions of databases with access by subject, database producer and/or vendor.

Kruzas, Anthony T., and John Schmittroth, Jr., eds. *Encyclopedia of Information Systems and Services.* Detroit, Mich.: Gale Research Co., 1981.

Contains listings and descriptions of databases, database producers and publishers, telecommunications networks, fee-based information services, and other information systems and services.

Williams, Martha E., ed. *Computer-Readable Data Bases: A Directory and Data Sourcebook.* White Plains, N.Y.: Knowledge Industry Publications, Inc., 1982.

Originally published in 1979 by the American Society for Information Sciences. Gives very detailed descriptions of individual databases.

Periodicals

Many periodicals, including those previously cited, are useful for online searchers and search service managers, but there are three periodicals that should be required reading in the library/information center online search service:

Online: The Magazine of Online Information Systems. Online, 11 Tannery Lane, Weston, Conn. 06883.

Online Review, Learned Information, Inc., The Anderson House, Stokes Road, Medford, N.J. 08055.

Database, Online, 11 Tannery Lane, Weston, Conn. 06883.

In addition, *RQ* now contains 150–200 word reviews of individual databases in the section of the periodical called "Sources."

The Evaluation of Information Retrieval Services: A Selected Bibliography

Compiled by RASD-MARS COMMITTEE ON
MEASUREMENT AND EVALUATION OF SERVICE

Blair, John C., Jr. "Measurement and Evaluation of Online Services." Chap. 5 in *The Library and Information Manager's Guide to Online Services*, Ryan E. Hoover, ed. White Plains, N.Y.: Knowledge Industry Publications, 1980. Thorough discussion of measurement and evaluation including record keeping, in-house evaluation of searches, and user evaluation of the service. Numerous charts and forms are included with twenty-eight references.

Cooper, William S. "On Selecting a Measure of Retrieval Effectiveness." *Journal of the American Society for Information Science* 24, no. 2 (March–April 1973): 87–100; 24, no. 6 (November 1973): 413–24.
Cooper proposes a "naive" but thought-provoking evaluation scheme that concentrates on relevance and user satisfaction. Article includes ten references.

Fosdick, Howard. "An SDC-based On-line Search Service." *Special Libraries* 68, no. 9 (September 1977): 305–12.
The results of a study of patron reaction to an online service in an engineering research laboratory. The author lists many of the survey questions and analyzes the responses to each of them. Text includes thirty-five references.

Hitchingham, Eileen E. "Selecting Measures Applicable to Evaluation of Online Literature Searching." *Drexel Library Quarterly* 13, no. 3 (July 1977): 52–67.
A theoretical discussion of the need for evaluation and an explanation of which measures should be taken. Hitchingham's concern is whether resources devoted to online searching are being effectively used. Article includes thirty-nine references.

King, Donald W., and Edward C. Bryant. *The Evaluation of Information Services and Products.* Washington, D.C.: Information Resources Press, 1971. pp. 237–48.

MARS Measurement and Evaluation of Service Committee, Reference and Adult Services Division, American Library Association: Sarah Pritchard (chair), Linda Friend, Marcia J. Sprules, Linda R. Oppenheim, Judith Bernstein, Richard Blood, Signe Larson, David Pilachowski, Elsie Ceruitti.

Of particular interest is the description of different modes of data collection in user surveys and the advantages and disadvantages of each.

Ladendorf, Janice. "Information Service Evaluation: The Gap between the Ideal and the Possible. *Special Libraries* 64, no. 7 (July 1979): 273–79.

Ladendorf discusses the purpose of evaluation, the limitations of some techniques of measurement, and decries the inadequacies in comparable library statistics. Article includes twenty-two references.

Lancaster, F. W. *The Measurement and Evaluation of Library Services.* Washington, D.C.: Information Resources Press, 1977.

The chapter entitled "Evaluation of Literature Searching and Information Retrieval" concentrates on the criteria by which users evaluate information services. "Library Surveys" examines the purpose, history, and literature of library surveys. Extensive bibliographies.

Penniman, W. D., and W. D. Dominick. "Monitoring and Evaluation of Online Information System Usage." *Information Processing and Management* 16, no. 1 (1980): 17–35.

The state-of-the-art on monitoring online system usage and on the techniques of data gathering for user evaluation. These techniques are evaluated and recommendations made for further study. Article includes fifty-four references.

Soergel, Dagobert. "Is User Satisfaction a Hobgoblin?" *Journal of the American Society for Information Science* 27, no. 4 (July–August 1976): 256–59.

Written in response to Cooper's article, Soergel makes the case that recall, the percentage of all potentially useful citations that are actually retrieved, is an important consideration, although it may be difficult to determine. Article includes five references.

Swanson, Rowena W. "Performing Evaluation Studies in Information Science." *Journal of the American Society for Information Science* 26, no. 3 (May–June 1975): 140–56.

Swanson discusses various aspects important in evaluation methodology, criteria, selection of variables, reproducibility, and reliability. Tends to be theoretical. Article includes thirty-four references.

Tagliacozzo, Renata. "Estimating the Satisfaction of Information Users." *Medical Library Association Bulletin* 65, no. 2 (April 1977): 243–49.

A useful examination of a questionnaire sent to 1000 MEDLINE users. Major conclusions, based on over 900 responses, were that (1) distinction should be made between user's satisfaction with a particular search and the user's feelings about the information retrieval system in general and (2) that it is potentially dangerous to isolate any one response on a questionnaire from any other. Article includes eight references.

Tessier, Judith A., Wayne W. Crouch, and Pauline Atherton. "New Measures of User Satisfaction with Computer-based Literature Searches." *Special Libraries* 68, no. 11 (November 1977): 383–89.

The authors describe four aspects of user satisfaction and the implied underlying assumption in user satisfaction measurement. They suggest additional measures of user satisfaction and ways to gain insights into them. Article includes twenty references.

Vickery, A., and A. Batten. *Large-scale Evaluation Study of On-line and Batch Computer Information Services.* London: Library Resource Co-ordinating Committee, University of London, 1978.

A report of a systematic study of online and batch information services at the University of London Library, including analysis of the problem, the design of procedure, the record of data collection, interpretation of results, and copies of forms used.

Warden, Carolyn L. "User Evaluation of a Corporate Library On-line Search Service." *Special Libraries* 72, no. 2 (April 1981): 113–17.

The results of the user study are discussed. A copy of the evaluation questionnaire is included. Article includes nine references.

Contributors

Janet Bruman, Senior Coordinator, Online Services
California Library Authority for Systems and Services
San Jose, California

MaryJane S. Cochrane, Head, Computer-Assisted Research Service
McKeldin Library
University of Maryland
College Park, Maryland

M. R. Dustin, Reference Coordinator, Minnesota Interlibrary
  Telecommunications Exchange
University of Minnesota
Minneapolis, Minnesota

John Edward Evans, Head of Reference/Microforms
Memphis State University Libraries
Memphis, Tennessee

Carol Hansen Fenichel, Director of Library Services
Joseph W. England Library
Philadelphia College of Pharmacy and Science
Philadelphia, Pennsylvania

Gertrude Foreman, Head of Reference
Bio-Medical Library
University of Minnesota
Minneapolis, Minnesota

Nancy E. Grimes, Online Retrieval System Analyst
Customer Services
DIALOG Information Services, Inc.
Palo Alto, California

Contributors

Randolph E. Hock, Boston District Manager
DIALOG Information Services, Inc.
Cambridge, Massachusetts

Sara D. Knapp, Coordinator of Information Retrieval
University Libraries
State University of New York at Albany

Gayle McKinney, Online Search Services Coordinator
Reference Department
Pullen Library
Georgia State University
Atlanta, Georgia

Lawrence R. Maxted
Microcomputer Consultant
Erie, Pennsylvania

James Rettig, Reference Librarian
Roesch Library
University of Dayton
Dayton, Ohio

Kristine Salomon, Social Science Reference Librarian
Reference Department
University Library
University of Nebraska
Omaha, Nebraska

Emelie J. Shroder, Chief
Business, Science and Technology Division
Chicago Public Library
Chicago, Illinois

Jane I. Thesing, Assistant Reference Librarian
Thomas Cooper Library
University of South Carolina
Columbia, South Carolina

Carol M. Tobin, Reference Librarian
General Reference Division
Princeton University Library
Princeton, New Jersey

Peter G. Watson, Data Services Coordinator
Meriam Library
California State University
Chico, California

Rebecca Whitaker, Information Retrieval Specialist
Indiana Cooperative Library Services Authority
Indianapolis, Indiana
and chair-elect, Machine-Assisted Reference Section
Reference and Adult Services Division
American Library Association

EDITOR

James J. Maloney, Head, Information Retrieval Services Division
Bibliographical Center for Research
Denver, Colorado
and chair, 1982 Program Committee
Machine-Assisted Reference Section
Reference and Adult Services Division
American Library Association